GC/WORKS/1 TWO STAGE DESIGN & BUILD (1999) GENERAL CONDITIONS

London: The Stationery Office

This document may be cited as - GC/Works/1 Two Stage Design & Build (1999)

Published with the permission of the Property Advisers to the Civil Estate on behalf of the Controller of Her Majesty's Stationery Office.

First published 1999

ISBN 0 11 702326 4

CONTENTS

INTRODUCTION AND NOTICES

THE GC/WORKS FAMILY OF CONTRACTS

GC/Works/1 (1998) is a new edition of the standard Government forms of contract for major UK building and civil engineering works, replacing the following three standard forms of contract -

- *GC/Works/1 (Edition 3) Lump Sum With Quantities, December 1989, revised 1990.*
- *GC/Works/1 (Edition 3) Lump Sum Without Quantities, August 1991.*
- *GC/Works/1 (Edition 3) Single Stage Design & Build Version, July 1993.*

GC/Works/1 (1998) is published in four volumes -

- *GC/Works/1 With Quantities (1998).*
- *GC/Works/1 Without Quantities (1998).*
- *GC/Works/1 Single Stage Design & Build (1998).*
- *GC/Works/1 Model Forms and Commentary (1998).*

These have now been joined by *GC/Works/1 Two Stage Design & Build (1999)* the *GC/Works/1 With Quantities Construction Management Trade Contract (1999)* and the *GC/Works/1 Without Quantities Construction Management Trade Contract (1999).*

PROCUREMENT METHODS

GC/Works/1 With Quantities (1998) is for use with Bills of Quantities, where all or most of the quantities are firm and not subject to re-measurement, giving a lump sum contract subject to adjustment for variations ordered.

GC/Works/1 Without Quantities (1998) is for use when lump sum tenders are to be invited on the basis of Specification and Drawings only, without Bills of Quantities, but supported by a Schedule of Rates prepared by the Contractor in order to value variations ordered.

GC/Works/1 Single Stage Design & Build (1998) is also a lump form of contract, which is intended to support a single stage tendering procedure, without a separate design stage. The form of contract is sufficiently flexible to allow for varying amounts of design input from the Contractor. The Employer provides Employer's Requirements, to which the Contractor responds with Contractor's Proposals, and the Contractor develops the detailed design on the basis outlined in those documents.

GC/Works/1 Two Stage Design & Build (1999) is intended to support a two stage procedure, with a separate design stage. It is also a lump sum form of contract, but the lump sum is arrived at in two stages, as explained in the Commentary section of this volume.

SUPPORTING DOCUMENTATION

GC/Works/1 Model Forms and Commentary (1998) contains forms which may be used with any of the three *GC/Works/1 (1998)* set of General Conditions.

The model forms required for use *GC/Works/1 Two Stage Design & Build (1999)* are contained in the disk which accompanies this volume. Guidance notes on each of the conditions in this contract can be found in *GC/Works/1 Model Forms and Commentary (1998)*. Guidance notes on the conditions specific to *GC/Works/1 Two Stage Design and Build (1999)* can be found at the back of this volume.

DISCLAIMER

All parties must rely exclusively upon their own skill and judgement, or upon those of their advisers, when making use of this document. Neither the Crown, nor Pinsent Curtis, nor any other contributor, assumes any liability to anyone for any loss or damage caused by any error or omission, whether such error or omission is the result of negligence or any other cause. Any and all such liability is disclaimed.

SUPPLEMENTARY CONDITIONS

Conditions, such as a variation of price condition, may be added to supplement the printed Conditions of Contract according to circumstances. They are incorporated into the Conditions of Contract by listing in the Abstract of Particulars.

If the parties incorporate Supplementary Conditions into the Conditions of Contract by listing in the Abstract of Particulars, such Supplementary Conditions will, as provided by the Abstract, prevail over the printed Conditions.

FURTHER GUIDANCE

Further guidance on this or any other document in the GC/Works suite of contracts can be obtained from:

Robert Pilling MCIPS
Procurement Contracts Adviser
PACE
Central Advice Unit
Trevelyan House
Great Peter Street
London SW1P 2BY

Telephone: 0171 217 2721

Fax: 0171 271 2715

Email: rwpsc@property.gov.uk

GENERAL CONDITIONS OF CONTRACT GC/WORKS/1 TWO STAGE DESIGN & BUILD (1999)

CONTRACT DOCUMENTATION, INFORMATION AND STAFF

1
Definitions, etc.

(1) In the Contract, unless the context otherwise requires -

'the Abstract of Particulars' means the document so headed included in the Contract;

'the Accepted Risks' means the risks of -

(a) pressure waves caused by the speed of aircraft or other aerial devices;

(b) ionising radiations or contamination by radioactivity from any nuclear fuel or from nuclear waste from the combustion of nuclear fuel;

(c) the radioactive, toxic, explosive or other hazardous properties of any explosive nuclear assembly (including any nuclear component); and

(d) war, invasion, act of foreign enemy, hostilities (whether or not war has been declared), civil war, rebellion, insurrection, or military or usurped power;

'CDM Regulations' means the Construction (Design and Management) Regulations 1994 or, if the Works are in Northern Ireland, the Construction (Design and Management) Regulations (Northern Ireland) 1995;

'Company' means and includes any body corporate;

'the Contract' means the -

(a) Contract Agreement (if any);

(b) Conditions of Contract (namely, these Conditions and any Supplementary Conditions and Annexes prescribed by the Abstract of Particulars);

(c) Abstract of Particulars;

(d) Employer's Requirements;

(e) Contractor's Proposals;

(f) Programme;

(g) Pricing Document;

(h) tender; and

(i) Employer's written acceptance;

'the Contract Agreement' means the formal agreement (if any) executed by the Employer and the Contractor recording the terms of the Contract;

'the Contract Sum' means the sum accepted by the Employer when notifying the Contractor when he may take possession of the Site, or those parts of the Site defined in the Contract, pursuant to Condition 34 (Commencement and completion);

'the Contractor' means the person or persons whose tender is accepted by the Employer and his or their legal personal representatives or permitted assignees;

'the Contractor's Proposals' means the statements, drawings, and other documents included in the Contract which define how the Contractor is to implement the Employer's Requirements;

'the Date or Dates for Completion' means the date or dates ascertained in accordance with the Abstract of Particulars or, where extensions of time have been awarded or acceleration agreed, the date as adjusted by such extensions or agreements;

'Days' means calendar days;

'the Design' means the complete design of the Works in accordance with the Employer's Requirements, and shall include (without limitation) the Contractor's Proposals and all Design Documents necessary for the design and execution of the Works in accordance with the Contract;

'Design Document' means any plan, sketch, drawing (including, without limitation, setting-out drawings), calculation, specification, and any other document whatsoever in any medium prepared by or on behalf of the Contractor in the performance of the Contract (or in preparation for the Contract) for, or in connection with, the Design (including, without limitation, any such document included in the Contractor's Proposals);

'the Design Fee' means the part of the Contract Sum accepted by the Employer, when awarding the Contract, and identified as such in the Pricing Document, as payable to the Contractor for carrying out the design process required under Condition 10B (Design process and Contract Sum);

'Design Process Event' means the determination of the Contract Sum in accordance with Condition 10B (Design process and Contract Sum), and the happening of the event specified in the Abstract of Particulars and shown in the Programme for the purposes of Condition 10B (Design process and Contract Sum) (and, if no such event is specified in the Abstract of Particulars, that event shall be the completion of the Design and the signing and dating of all Design Documents by the PM under Condition 10A(2) (Design Documents));

'the Employer' means the Employer in the Contract as named in the Abstract of Particulars and any permitted assignees;

'the Employer's Requirements' means the statements, drawings, and other documents included in the Contract which define the Employer's requirements;

'the Final Account' means the document prepared by the QS showing the calculation of the Final Sum and agreed or otherwise settled under Condition 49 (Final Account);

'the Final Sum' means the amount payable under the Contract by the Employer to the Contractor for the full and entire execution and completion of the Works;

'Group' means and includes a company and every holding company of that company for the time being, and every subsidiary for the time being of every such holding company, and the terms 'Employer's Group' and 'Contractor's Group' shall be interpreted accordingly; but, while the Employer is a Minister of the Crown, a government department or other Crown agency or authority, the term 'Employer's Group' shall also include all other Ministers of the Crown, government departments and Crown agencies and authorities;

'the Health and Safety Plan' means, where it is stated in the Abstract of Particulars that all the CDM Regulations apply, the plan provided to the Principal Contractor and developed by him to comply with Regulation 15 of the CDM Regulations and, for the purpose of Regulation 10 of the CDM Regulations, received by the Employer before any construction work under the Contract has started; and any further development of that plan by the Principal Contractor during the progress of the Works;

'Holding company' shall have the meaning given in Section 736 of the Companies Act 1985, as substituted by Section 144 of the Companies Act 1989;

'Instruction' means any instruction given in accordance with Condition 40 (PM's Instructions), and, except where expressly stated to the contrary, includes any Variation Instruction;

'the Maintenance Period' means the period, or any of the periods, specified in the Abstract of Particulars for the rectification of defects in accordance with Condition 21 (Defects in Maintenance Periods);

'Milestone' means the completion of each of the phases of the Works described in the Milestone Payment Chart (if any);

'Milestone Payment Chart' means the chart or table (if any) included with the invitation to tender or agreed by the Employer which specifies the amounts of the advance payments to be made to the Contractor during the performance of the Works upon achievement of the Milestones;

'Planning Supervisor' means the PM, or any other person named in the Abstract of Particulars as Planning Supervisor, or such other person as may be appointed in that capacity for the time being by the Employer pursuant to Regulation 6(5) of the CDM Regulations;

'Pricing Document' means the document included in the Contract in explanation and support of the Contract Sum;

'Principal Contractor' means the Contractor, or such other contractor or organisation as may be appointed as the Principal Contractor for the time being by the Employer pursuant to Regulation 6(5) of the CDM Regulations;

'the PM' means the Project Manager who is the person employed in that capacity named in the Abstract of Particulars and appointed by the Employer to act on his behalf in carrying out those duties described in the Contract (subject to the exclusions set out in the Abstract of Particulars), or such other person as may be appointed in that capacity for the time being by the Employer;

'the Programme' means the programme submitted prior to acceptance of the tender and agreed at that time by the Employer, as it may be amended from time to time;

'the QS' means the Quantity Surveyor appointed for the time being by the Employer;

'Retention Payment' shall have the meaning given in Condition 48A (Retention payment bond);

'Section' means any part of the Works specified as a Section in the Abstract of Particulars and which has a specified Date for Completion;

'the Site' means the land or place described in the Contract, together with such other land or places as may be allotted or agreed by the parties from time to time, for the purpose of carrying out the Contract;

'the Stage Payment Chart' means the chart, table or graph (if any) included with the invitation to tender or agreed by the Employer which specifies the periods and the percentage amount of the advance payments to be made to the Contractor during the performance of the Works;

'Subsidiary' shall have the meaning given in Section 736 of the Companies Act 1985, as substituted by Section 144 of the Companies Act 1989;

'Things' comprise 'Things for incorporation', which means goods and materials intended to form part of the completed Works, and 'Things not for incorporation' which means goods and materials provided or used to facilitate execution of the Works but not for incorporation in them;

'Unforeseeable Ground Conditions' means ground conditions certified by the PM in accordance with Condition 7 (Conditions affecting Works);

'Variation Instruction' ('VI') means any Instruction which makes any alteration or addition to, or omission from, the Works or any change in the design, quality or quantity of the Works, as a consequence of any change in the Employer's Requirements; and

'the Works' means the works described or shown in the Employer's Requirements, Contractor's Proposals, Pricing Document, and/or Design, including (without limitation) the Design itself and all modified or additional works to be executed under the Contract.

(2) The headings to these Conditions shall not affect their interpretation. Any reference to legislation shall be deemed to include a reference to any amendment or re-enactment thereof for the time being in force. Words in the singular include the plural, and *vice versa*. Words in the masculine include feminine and neuter.

(3) Unless otherwise provided by the Contract, certificates of the PM shall be issued to the Contractor. Any notices or other documents required or authorised to be served in pursuance of the Contract shall be in writing. They may be given to the Contractor by delivery to his agent or may be posted to the registered office or last known place of business of the Contractor. They may be given to the Employer by delivery to the PM or may be posted to the registered office or last known place of business of the Employer. A postal notice shall be deemed to have been served on the date when in the ordinary course of post it would have been delivered.

(4) (a) Except in relation to Condition 36 (Extensions of time), any period of time in these Conditions within which the Employer, the Contractor, PM or QS, is to take action or decide anything may be extended by agreement, notwithstanding that the period of time has expired.

(b) If the Works are in England, Wales or Scotland, then for the purposes of this Contract, and for all other purposes, periods of time shall include Saturdays and Sundays, but shall be reckoned as follows -

(i) where an act is required to be done within a specified period after or from a specified date, the period begins immediately after that date; and

(ii) where the period would include Christmas Day, Good Friday or a day which under the Banking and Financial Dealings Act 1971 is a bank holiday in England and Wales, or as the case may be, in Scotland, that day shall be excluded.

(c) If the Works are in Northern Ireland, then for the purposes of this Contract, and for all other purposes, subject to Condition 50(4) (Certifying payments) periods of time shall include Saturdays and Sundays, but shall be reckoned as follows -

(i) where a period of time is expressed to begin on, or to be reckoned from, a particular day, that day shall not be included in the period;

(ii) where the time limited for the doing of anything expires or falls upon a Sunday or a public holiday, the time so limited shall extend to, and the thing may be done, on the first following day that is not a Sunday or a public holiday; and

(iii) the expression 'public holiday' shall include Christmas Day, Good Friday, any bank holiday appointed by or under any statutory provision, and any day appointed for public thanksgiving or mourning.

1A

Fair dealing and teamworking

(1) The Employer and the Contractor shall deal fairly, in good faith and in mutual co-operation, with one another, and the Contractor shall deal fairly, in good faith and in mutual co-operation, with all his subcontractors and suppliers.

(2) Both parties accept that a co-operative and open relationship is needed for success, and that teamwork will achieve this. The project team for this purpose shall include, but shall not be limited to, the PM and his representatives; the QS; the Contractor's agent; and major subcontractors and suppliers engaged on the Works from time to time.

(3) As soon as possible, the PM shall call a project team meeting and agree a programme of regular meetings with the Contractor. Either the PM or the Contractor may also call additional meetings of the team, and if the other agrees may invite any other person needed for an effective meeting. The PM and the Contractor shall use the meetings jointly to develop proposals for reducing costs by solutions that will, so far as possible, be to the benefit of all affected by potential or actual problems. The meetings shall also consider the issues of advance warning of anything which might raise costs or harm final quality of the Works.

2

Contract documents

(1) In case of discrepancy between the Supplementary Conditions and Annexes prescribed by the Abstract of Particulars and these Conditions, the former shall prevail. In case of discrepancy between the Conditions of Contract and other documents forming part of the Contract, the Conditions of Contract shall prevail.

(2) In case of discrepancy between the Employer's Requirements and either the Contractor's Proposals or the Pricing Document, the Employer's Requirements shall prevail, without adjustment to the Contract Sum.

(3) In case of any discrepancy within the Employer's Requirements, or any discrepancy between the Employer's Requirements and any statutory requirement, building regulation or planning permission, the PM shall resolve the discrepancy, which shall be treated as a change in the Employer's Requirements.

(4) In case of any discrepancy within or between either the Contractor's Proposals and/or the Pricing Document, the PM may choose between discrepant items or agree any proposal from the Contractor to resolve the discrepancy, without adjustment to the Contract Sum.

(5) If the Contractor discovers any of the discrepancies referred to in paragraphs (2), (3) and (4), he shall immediately notify the PM, informing him of the discrepancy.

(6) The PM shall provide free to the Contractor as soon as reasonably practicable a copy of the Employer's Requirements, in a form which the PM considers suitable for reproduction. The Contractor shall keep a copy of the Employer's Requirements and the Contractor's Proposals on Site, to which the PM or his representatives shall have access at all reasonable times.

(7) Figured dimensions on all drawings shall be followed in preference to the scale.

3

Pricing Document

(1) There shall be no rectification of any errors or omissions in the Pricing Document (including, without limitation, errors or omissions in respect of wrong quantities, wrong estimates of prices, calculations or computations).

(2) The Contractor shall keep a copy of the Pricing Document on the Site, to which the PM or his representatives shall have access at all reasonable times.

4 Delegations and representatives

(1) All decisions to be made by or on behalf of the Employer under the Contract shall be communicated to the Contractor by the PM. The PM shall be deemed to be authorised to act generally for the Employer, subject to any excluded matters set out in the Abstract of Particulars. In relation to those excluded matters, the person or persons authorised to act for the Employer are identified in the Abstract of Particulars, but decisions in respect of excluded matters shall nevertheless be communicated to the Contractor by the PM.

(2) The PM and QS may expressly delegate in writing to named representatives any of their powers and duties. Where a Clerk of Works or Resident Engineer is appointed he shall exercise the powers of the PM under Condition 31 (Quality) and such other powers as the PM may delegate to him.

(3) The appointment of representatives by the Employer, PM or QS shall not prevent them from subsequently exercising directly any of the powers and duties conferred under the Contract.

(4) The Contractor shall as soon as possible be notified of all powers and duties delegated, and of the names of representatives, and of any subsequent changes.

5 Contractor's agent

The Contractor shall employ a competent agent to supervise the execution of the Works. Except when required to attend at the office of the PM, or when reasonably absent from the Site for other reasons, the agent shall be in attendance at the Site during all working hours. When the agent is not in attendance at the Site, the Contractor will so notify the PM, stating the period of and reason for the absence, together with the name of the agent's authorised deputy.

6 Contractor's employees

(1) The PM may at any time require the Contractor immediately to cease to employ in connection with the Contract any person, including the Contractor's agent, whose continued employment is in the opinion of the PM undesirable. The Contractor shall replace any such person with a suitably qualified person.

(2) Other than for causes outside his control, the Contractor shall not make changes in personnel named in his tender in connection with the Contract without the prior approval of the PM.

GENERAL OBLIGATIONS

7

Conditions affecting Works

(1) The Contractor shall in relation to the Site be deemed to have satisfied himself as to -

- (a) the existing roads, railways and other means of communication with or access to it;
- (b) its contours and boundaries;
- (c) the risk of damage by reason of any work to any property adjacent to the Works and injury to occupiers of that property;
- (d) the nature of the soil and material (whether natural or otherwise) to be excavated;
- (e) the conditions under which the Works will have to be carried out, including precautions to prevent nuisance and pollution;
- (f) the supply of and conditions affecting labour necessary to carry out the Works;
- (g) the facilities for obtaining any Things whether or not for incorporation; and
- (h) any other matters or information affecting or likely to affect the execution of, or price tendered for, the Works.

(2) (Not used)

(3) If, during the execution of the Works, the Contractor becomes aware of ground conditions (excluding those caused by weather but including artificial obstructions) which he did not know of, and which he could not reasonably have foreseen having regard to any information which he had or ought reasonably to have ascertained, he shall, as a condition precedent to any right or remedy in respect of such conditions, by notice immediately -

- (i) inform the PM of those conditions; and
- (ii) state the measures which he proposes to take to deal with them.

(4) If the PM agrees that the ground conditions specified in a notice under paragraph (3) could not reasonably have been foreseen by the Contractor having regard to any information he should have had in accordance with that paragraph and paragraph (1), he shall certify those conditions to be Unforeseeable Ground Conditions. The PM shall notify the Contractor of his decision.

(5) If, after ground conditions specified in a notice under paragraph (3) have been or should have been certified as Unforeseeable Ground Conditions in accordance with this Condition, and as a result of such Unforeseeable Ground Conditions, the Contractor in executing the Works properly carries out or omits any work which he would not otherwise have carried out or omitted, then, without prejudice to any Instruction given by the PM, the value of the work carried out or omitted shall be ascertained in accordance with Condition 42 (Valuation of Variation Instructions) and the Contract Sum shall be increased or decreased accordingly.

(6) No claim by the Contractor for additional payment will be allowed because he has misunderstood or misinterpreted anything mentioned in paragraph (1). The Contractor shall not be released from any risks or obligations imposed on, or undertaken by, him under the Contract for this reason, or because he did not or could not foresee any matter which might affect or have affected the execution of the Works.

8
Insurance

(1) A party required under this Condition to effect or maintain insurance is called in this Condition 'the insuring party' but, while the Employer is a Minister of the Crown, a government department or other Crown agency or authority, the Employer shall be deemed not to be the insuring party.

(2) The Contractor shall by such existing or new policies as he sees fit effect and maintain from the time he takes possession of the Site or any part of the Site or from the time he commences the execution of the Works (if earlier) to the expiration of the last Maintenance Period to expire, employer's liability insurance in respect of persons in his employment, appropriate to the nature of the Works. Such insurance shall comply with the Employer's Liability (Compulsory Insurance) Act 1969 (or, if the Works are in Northern Ireland, the Employer's Liability (Defective Equipment and Compulsory Insurance) (Northern Ireland) Order 1972)) and any subordinate legislation made thereunder, and shall be for the minimum amount of £10,000,000 (or such other minimum amount as may be stated in the Abstract of Particulars) for any one occurrence or series of occurrences arising out of one event.

Alternative A

(3) The Contractor shall by such existing or new policies as he sees fit effect and maintain for the same period -

(a) construction 'all risks' insurance in the joint names of the Employer and the Contractor against loss or damage to the Works and Things for which the Contractor is responsible under the terms of the Contract for the full reinstatement value thereof (including transit and off-Site risks) plus 15% (or such other percentage as may be stated in the Abstract of Particulars) for professional fees; and

(b) public liability insurance against legal liability for personal injury to any persons and loss or damage to property arising from or in connection with the Works, which is not covered by employer's liability insurance under paragraph (2) or by insurance against loss or damage to the Works and Things under subparagraph (a), for the minimum amount stated in the Abstract of Particulars, such public liability insurance to include a provision for indemnity to the Employer in respect of the Contractor's liability under Condition 19 (Loss or damage);

provided that the insurance which the Contractor is required to effect and maintain under this paragraph need not cover loss or damage caused by any Accepted Risk.

Alternative B

(3) The Contractor shall effect and maintain insurance in accordance with the Summary of Essential Insurance Requirements attached to the Abstract of Particulars.

Alternative C

(3) (a) While the Employer is a Minister of the Crown, a government department or other Crown agency or authority, the Employer shall not be required to effect or maintain any insurance, but shall assume the risks of loss or damage which should have been insured if paragraph (3)(b) applied, and, in the event of loss or damage arising, shall be responsible as if paragraph 3(b) applied and the Employer had failed to effect and maintain insurance as described therein, but the Employer shall not be responsible for any amounts in excess of amounts which should have been insured, or any liability authorised to be retained, or risks which would not have been insured or would have been excluded by the application of the terms, exceptions or conditions of any such insurance.

(b) While the Employer is not a Minister of the Crown, a government department or other Crown agency or authority, the Employer shall effect and maintain insurance in accordance with the Summary of Essential Insurance Requirements attached to the Abstract of Particulars, but shall not be responsible for any amounts in excess of amounts insured, or any retained liability, or risks not insured or excluded by the application of the terms, exceptions or conditions of any such insurance.

---

(4) The other party shall have the right to receive, on request, a copy of insurances required to be effected or maintained by the insuring party under this Condition. The insuring party shall within 21 Days of the acceptance of the tender, and also within 21 Days of any subsequent renewal or expiry date of relevant insurances, send to the other party a certificate in the form attached to the Abstract of Particulars from his insurer or broker attesting that insurance has been effected in accordance with the Contract.

(5) All insurances required to be effected or maintained by the insuring party under this Condition shall be with reputable insurers, to whom the other party has no reasonable objection, lawfully carrying on such insurance business in the United Kingdom, and upon customary and usual terms prevailing for the time being in the insurance market. The said terms and conditions shall not include any term or condition to the effect that any insured must discharge any liability before being entitled to recover from the insurers, or any other term or condition which might adversely affect the rights of any person to recover from the insurers pursuant to the Third Parties (Rights Against Insurers) Act 1930 or the Third Parties (Rights Against Insurers) Act (Northern Ireland) 1930.

(6) All insurances required to be effected or maintained under paragraph (3) (Alternatives B or C) (if applicable) -

(a) shall be in the joint names of the Employer, such other persons as the Employer may reasonably require (including, without limitation, the Employer's consultants, any persons who have entered or shall enter into an agreement for the provision of finance in connection with the Works, and any persons who have acquired or shall acquire any interest in or over the Works or any part thereof), the Contractor and all subcontractors; and

(b) shall extend to cover loss or damage which the Contractor is responsible for making good under Condition 21 (Defects in Maintenance Periods) or which occurs while the Contractor is making good defects in the Works under that Condition.

(7) If, without the approval of the other party, the insuring party fails to effect and maintain insurance he is required to effect and maintain under this Condition as described, or obtains a different policy of insurance, or fails to provide a copy of insurances or certificates in accordance with this Condition, the other party may, but is not required to, effect and maintain appropriate insurance cover and deduct the cost of doing so from any payment due to the insuring party under the Contract, or recover such sum from the insuring party as a debt.

(8) If -

(a) the Works involve the refurbishment, alteration or extension of existing structures; and/or

(b) a 'completed part' within the meaning of Condition 37 (Early possession) is certified by the PM as having been completed in accordance with the Contract;

the Employer shall bear the risks of fire, lightning, explosion, storm, tempest, flood, bursting or overflowing of water tanks, apparatus or pipes, earthquake, aircraft or other aerial devices or articles dropped therefrom, riot and civil commotion (including terrorism) in respect of loss or damage to the existing structures and contents for which the Employer is responsible, and in respect of loss or damage to the completed part from the date of its certification, except where Condition

19 (Loss or damage) applies to the relevant loss or damage and to the extent that the Contractor is not entitled to reimbursement under Condition 19(5).

(9) For the avoidance of doubt, it is agreed that nothing in this Condition shall relieve the Contractor from any of his obligations and liabilities under the Contract.

8A

Professional indemnity insurance for design

(1) The Contractor shall maintain professional indemnity insurance covering (*inter alia*) all liability hereunder in respect of defects or insufficiency in design, upon customary and usual terms and conditions prevailing for the time being in the insurance market, and with reputable insurers lawfully carrying on such insurance business in the United Kingdom (in an amount not less than that required by the Abstract of Particulars) for any one occurrence or series of occurrences arising out of any one event, for a period beginning now and ending 12 years (or such other period as is required by the Abstract of Particulars) after certification under Condition 39 (Certifying completion) of the completion of the Works or the last Section thereof in respect of which completion is certified, or the determination of the Contract for any reason whatsoever, including (without limitation) breach by the Employer, whichever is the earlier, provided always that such insurance is available at commercially reasonable rates. The said terms and conditions shall not include any term or condition to the effect that the Contractor must discharge any liability before being entitled to recover from the insurers, or any other term or condition which might adversely affect the rights of any person to recover from the insurers pursuant to the Third Parties (Rights Against Insurers) Act 1930 or the Third Parties (Rights Against Insurers) Act (Northern Ireland) 1930. The Contractor shall not, without the prior approval in writing of the Employer, settle or compromise with the insurers any claim which the Contractor may have against the insurers and which relates to a claim by the Employer against the Contractor, or by any act or omission lose or prejudice the Contractor's right to make or proceed with such a claim against the insurers.

(2) The Contractor shall immediately inform the Employer if such insurance ceases to be available at rates that the Contractor considers to be commercially reasonable. Any increased or additional premium required by insurers by reason of the Contractor's own claims record, or other acts, omissions, matters or things particular to the Contractor shall be deemed to be within commercially reasonable rates.

(3) The Contractor shall fully co-operate with any measures reasonably required by the Employer, including (without limitation) completing any proposals for insurance and associated documents, maintaining such insurance at rates above commercially reasonable rates if the Employer undertakes in writing to reimburse the Contractor in respect of the net cost of such insurance to the Contractor above commercially reasonable rates or, if the Employer effects such insurance at rates at or above commercially reasonable rates, reimbursing the Employer in respect of what the net cost of such insurance to the Employer would have been at commercially reasonable rates.

(4) As and when reasonably required to do so by the Employer, the Contractor shall produce for inspection documentary evidence (including, if required by the Employer, the originals of the relevant insurance document) that his professional indemnity insurance is being maintained.

(5) The above obligations in respect of professional indemnity insurance shall continue notwithstanding determination of the Contract for any reason whatsoever, including (without limitation) breach by the Employer.

9

Setting out and 'as built' drawings

(1) The Contractor shall set out the Works, and shall provide all the instruments, profiles, templates and rods for that purpose. The Contractor shall be solely responsible for the correctness of the setting out.

(2) The Contractor shall provide, fix and be responsible for the maintenance of all stakes, templates, profiles, level marks, points and any other setting out apparatus which is required. He shall take all necessary precautions to prevent their removal, alteration or disturbance and shall be liable for the consequences of their removal, alteration or disturbance and for their efficient reinstatement.

(3) Not later than 14 Days after each date certified by the PM under Condition 39 (Certifying completion) as the date when the Works, or any Section, or any completed part within the meaning of Condition 37 (Early possession) are or is completed in accordance with the Contract, the Contractor shall, as a condition precedent to receiving any further certification or payment under the Contract, supply free to the Employer full drawings, documents and information as necessary to show and describe the Works, Section or completed part 'as built', together with such other information necessary to explain the operation and maintenance of the same. Such drawings, documents and information shall be supplied in three copies, or in such other number as is stated in the Abstract of Particulars, and shall be in the form and medium stated in the Employer's Requirements or Instructed by the PM.

10

Design

(1) The Contractor has undertaken and shall undertake, and is and shall be responsible for, the Design. He shall prepare all the Design Documents required for the complete and efficient execution of the Works in accordance with the Contract. The Contractor warrants that the Contractor's Proposals, all the Design Documents, and the Design, have been and shall be prepared with all reasonable skill, care and diligence. The Contractor further warrants that the Design and the Works meet, and shall continue to meet, the requirements of any relevant planning permission and the Building Regulations and comply, and shall continue to comply, with all other relevant statutory requirements. If, after the formation of the Contract, there is any change in any such requirements which affects the Works, then such change will be treated as a change in the Employer's Requirements.

Alternative A

(2) The Contractor's liability to the Employer in respect of any defect or insufficiency in the Design shall be the same as would have applied to an architect or other appropriate professional designer who had held himself out as competent to take on the Design and who had acted independently under a separate contract with the Employer and supplied the Design for, or in connection with, works to be carried out and completed by a contractor not being the supplier of the Design.

Alternative B

(2) The Contractor warrants to the Employer that the Works will be fit for their purposes, as made known to the Contractor by the Contract.

---

(3) The Contractor's liability under this Condition shall not be affected by any warranty that the Employer may obtain from any subcontractor.

(4) Subject to Condition 10A(3) (Design Documents), the copyright in all documents provided by the Contractor or any subcontractor in connection with the Works shall remain vested in the copyright

owner (or as may be otherwise provided by the relevant subcontract), but the Employer and its appointee shall have, and the Contractor warrants and shall procure that the Employer and its appointee shall have, a licence to copy and use such documents, and to reproduce the designs contained in them, for any purpose related to the Works including, without limitation, the construction, completion, maintenance, letting, promotion, advertisement, reinstatement, repair and extension of the Works. The Contractor shall, at any time and on one or more occasions, if the Employer so requests and undertakes in writing to pay the Contractor's reasonable copying charges, promptly supply the Employer with conveniently reproducible copies of all such documents.

10A

Design Documents

(1) To demonstrate compliance with the Employer's Requirements, the Contractor shall ensure that relevant work will be the subject of a Design Document. The Contractor shall provide free to the PM two copies (or such other number as is stated in the Abstract of Particulars) of each Design Document before commencing the execution of the work for which the Design Document has been prepared. Each Design Document shall be in the form and medium stated in the Employer's Requirements or Instructed by the PM. The Contractor shall respond promptly to any questions about the Design Documents raised by the PM.

(2) The Contractor shall not commence any work in accordance with any Design Document until he has ascertained that -

(a) the PM has examined the Design Document and has confirmed in writing that he does not intend to raise any questions thereon; or

(b) the PM has confirmed in writing that the questions he has raised about the Design Document have been answered to his satisfaction; and

upon completion of the procedures in (a) or (b) above the PM shall sign and date each copy of the Design Document, and return one copy to the Contractor.

(3) Notwithstanding Condition 10(4) (Design), the copyright in the Design and in all Design Documents shall be the property of and shall vest in the Employer, and the Contractor so warrants.

(4) The Contractor shall keep one copy of each Design Document on the Site, to which the PM or his representatives shall have access at all reasonable times.

(5) Nothing in this Condition shall relieve the Contractor of any liability under the Contract for any defect in any Design Document, or for any inconsistency or lack of co-ordination between any Design Documents.

(6) In the context of paragraphs (1) and (2), the Contractor must submit the Design Documents for examination within a period which would reasonably allow the PM to raise queries, and for the Contractor to respond, before the date on which the Contractor intends to commence work. Design Documents shall be submitted in recorded, dated and logically linked batches.

(7) In case of any discrepancy between the Employer's Requirements and any Design Document, the Employer's Requirements shall prevail, without any adjustment to the Contract Sum.

(8) In case of discrepancy within or between Design Documents, the Contractor shall resolve the discrepancy to the satisfaction of the PM, without any adjustment to the Contract Sum.

(9) If the Contractor discovers any of the discrepancies referred to in paragraphs (7) and (8) he shall immediately notify the PM, informing him of the discrepancy.

10B

Design process and Contract Sum

(1) The Contractor shall carry out any site and/or soil investigations required by the Contract and proceed with the Design so that the Design Process Event shall be achieved in accordance with the Programme. Unless and until the Employer notifies the Contractor, pursuant to Condition 34 (Commencement and completion), when he may take possession of the Site or those parts of the Site defined in the Contract, the Contractor shall not execute any other part of the Works except as required by this Condition, unless so Instructed.

(2) As soon as reasonably practicable, and not later than 56 Days before the time shown in the Programme for the certification of the Design Process Event, the Contractor shall submit to the QS a written fully itemised lump sum quotation of the proposed Contract Sum, based as far as possible on the rates and prices contained within the Pricing Document. The Contractor shall include with his quotation such information as will enable the QS to evaluate the quotation, and shall also promptly submit to the QS any information he reasonably requires in explanation and support of the quotation.

(3) The Contract Sum shall be determined:-

(a) by acceptance by the PM of a lump sum quotation of the proposed Contract Sum prepared by the Contractor and submitted to the QS in accordance with paragraph (2), or such other lump sum as may be agreed by negotiation on such a quotation; or

(b) by the QS in accordance with paragraph (6).

(4) The PM shall notify the Contractor, as soon as reasonably practicable after the receipt of any such lump sum quotation, whether or not it is accepted or, if it is not acceptable, whether he is prepared to agree any other lump sum. If accepted, the aggregate amount specified in the quotation, or otherwise agreed between the Employer and the Contractor, shall be the Contract Sum.

(5) Should:-

(a) the Contractor fail to provide a lump sum quotation in accordance with paragraph (2); or

(b) the Employer and the Contractor fail to agree by negotiation;

then the PM shall instruct the QS to determine the Contract Sum.

(6) Where the QS is required to determine the Contract Sum he shall do so:-

(a) by measurement and valuation at the rates and prices in the Pricing Document; or

(b) if it is not possible to value as in subparagraph (a), then by measurement and valuation at rates and prices deduced or extrapolated from the rates and prices in the Pricing Document; or

(c) if it is not possible to value as in subparagraph (b), then by measurement and valuation as fair rates and prices, having regard to current market prices.

(7) The QS shall as soon as reasonably practicable notify the Contractor of his determination of the Contract Sum. Such determination shall, subject to paragraph (8), be subject to adjudication under Condition 59 (Adjudication) and arbitration under Condition 60 (Arbitration and choice of law).

(8) If the Contractor disagrees with the whole or part of the QS's determination of the Contract Sum, he shall, within 14 Days of the QS's notification under paragraph (7), give his reasons for disagreement and his own valuation. In any other case, and in all other respects, he shall be treated as having accepted the notification under paragraph (7), and the QS's determination of the Contract Sum.

(9) Any percentage or lump sum adjustments made in the pricing of the Pricing Document shall be deemed to be applicable to the determination of the Contract Sum.

(10) The PM shall certify the date when the Design Process Event is achieved.

11

Statutory notices and CDM Regulations

(1) (a) The Contractor shall give all notices required by any Act of Parliament or by any regulations or bye-laws made under any Act which may be required in connection with the Works. He shall pay any fees or charges required to be paid under any Act, regulations or bye-laws in respect of the Works and supply all drawings and plans required in connection with any notice.

(b) In this Condition the expression 'Act of Parliament' shall include an Act of the Scottish Parliament, an Act of the Parliament of Northern Ireland, a measure of the Northern Ireland Assembly and an Order in Council made under Section 1(3) of the Northern Ireland (Temporary Provisions) Act 1972 or under paragraph 1(1) of Schedule 1 to the Northern Ireland Act 1974; and the expressions 'regulation' and 'bye-laws' shall include subordinate legislation made under an Act of Parliament.

(2) The Contractor shall obtain the consent, permission or licence of any statutory undertakers, or any adjoining owners whose services or land may be affected by, or whose consent is necessary in connection with, the Works. The Contractor shall pay any licence fee or charge required in connection with any consent or licence.

(3) The Employer shall reimburse the amount of any fee or charge paid by the Contractor which he had properly incurred in accordance with paragraphs (1) or (2).

(4) Where the Contractor is and remains the Principal Contractor, he will meet all the duties of a Principal Contractor as set out in the CDM Regulations and he shall liaise with the Planning Supervisor and ensure that any employee, agent, subcontractor or supplier of his complies with the CDM Regulations and co-operates with the Planning Supervisor so as to enable the Planning Supervisor to carry out his duties under the CDM Regulations.

(5) Before commencing the Works, the Contractor shall submit to the PM the Health and Safety Plan containing the information required by Regulation 15 of the CDM Regulations and the Contractor shall not commence the Works until the PM has confirmed in writing to the Contractor that the Plan is of an appropriate standard. Such confirmation shall not relieve the Contractor of his liability under the Contract or the CDM Regulations. The Contractor shall notify the PM and the Planning Supervisor of all amendments to the Plan.

(6) Where the Employer, necessarily in order to comply with the CDM Regulations, appoints a successor to the Contractor as Principal Contractor, the Contractor will meet all reasonable costs to the Employer of any successor Principal Contractor. Notwithstanding Condition 36 (Extensions of time), no extension of time will be given by reason of the removal of the Contractor as Principal Contractor, and/or of the appointment of any successor Principal Contractor.

(7) Provided a reasonable period of written notice is given to the Contractor by the Planning Supervisor, the Contractor will provide, and will ensure that any subcontractor provides through him, such information as the Planning Supervisor needs for the preparation of a health and safety file as required by Regulations 14(d)(e) and (f) of the CDM Regulations. If the Contractor is not the Principal Contractor, such enquiries will be routed via the Principal Contractor.

12
Intellectual property rights

(1) The Contractor shall pay any royalty, licence fee or other expense for the supply or use of any patent, process, drawing, model, plan, invention or information used or necessary for, or in connection with, the Works.

(2) Where the use or supply of any patent, etc., in accordance with paragraph (1) gives rise to any claim or proceedings against the Employer, the Contractor shall reimburse the Employer any costs and expenses reasonably incurred by the Employer in respect of such claim or proceedings.

(3) The Employer shall reimburse the Contractor the amount of any royalty, etc., incurred in accordance with paragraph (1) which -

(a) was necessarily incurred in order to comply with a VI; and

(b) was not reasonably contemplated under the Contract.

13
Protection of Works

(1) The Contractor shall during the execution of the Works take all reasonable measures and precautions needed to take care of the Site and the Works, and shall have custody of all Things on the Site against loss or damage from fire and any other cause. The Contractor shall be solely responsible for and shall take all reasonable and proper steps for protecting, securing, lighting and watching all places on or about the Works and the Site which may be dangerous to his workpeople or to any other person.

(2) The Contractor shall comply with any statutory regulations (whether or not binding on the Crown) which govern the storage and use of all Things which are brought on to the Site in connection with the Works.

14
Nuisance and pollution

The Contractor shall take all reasonable precautions to prevent any nuisance or inconvenience to the owners, tenants or occupiers of any other property and to the general public, and shall secure the efficient protection of all streams and waterways against pollution.

15
Returns

The Contractor's agent shall provide the PM with a return, in the form the PM shall direct, of the number and description of workpeople and the plant employed each Day on the Works.

16
(Not used)

17
Covering work

The Contractor shall give reasonable notice to the PM whenever any work or Thing for incorporation is intended to be covered with earth or otherwise. In default of such notice, the Contractor shall, if required by the PM, uncover the work or Thing at his own expense.

18 Measurement

(1) The Contractor's representative shall, when required on reasonable notice by the QS, attend at the Works to take jointly with the QS measurements of the work executed. These measurements and any differences in relation to them shall be recorded in the manner required by the QS.

(2) The Contractor shall without extra charge provide the appliances and other things necessary for measuring the work.

(3) If the Contractor's representative fails to attend when required in accordance with this Condition, the QS may proceed to take any measurements.

19 Loss or damage

(1) This Condition applies to any loss or damage which arises out of, or is in any way connected with, the execution or purported execution of the Contract.

(2) The Contractor shall without delay and at his own cost reinstate, replace or make good to the satisfaction of the Employer or, if the Employer agrees, compensate the Employer for, any loss or damage.

(3) Where a claim is made, or proceedings are brought against the Employer in respect of any loss or damage, the Contractor shall reimburse the Employer any costs or expenses which the Employer may reasonably incur in dealing with, or in settling, that claim or those proceedings.

(4) The Employer shall notify the Contactor as soon as possible of any claim made, or proceedings brought, against the Employer in respect of any loss or damage.

(5) The Employer shall reimburse the Contractor for any costs or expenses which the Contractor incurs in accordance with paragraphs (2) and (3) to the extent that the loss or damage is caused by -

- (a) the neglect or default of the Employer or of any other contractor or agent of the Employer;
- (b) any Accepted Risk or Unforeseeable Ground Conditions; or
- (c) any other circumstances which are outside the control of the Contractor or of any of his subcontractors or suppliers and which could not have been reasonably contemplated under the Contract; provided that this subparagraph shall not apply where the loss or damage is loss or damage falling within subparagraph 6(c).

(6) In this Condition loss or damage includes -

- (a) loss or damage to property;
- (b) personal injury to, or the sickness or death of, any person;
- (c) loss or damage to the Works or to any Things on the Site; and
- (d) loss of profits or loss of use suffered because of any loss or damage.

20 (Not used)

21 Defects in Maintenance Periods

(1) The Contractor shall without delay make good at his own cost any defects in the Works, resulting from what the Employer considers to be default by the Contractor or his agent or subcontractors or suppliers, which appear during the relevant Maintenance Period.

(2) After completion of the remedial works by the Contractor, the Employer shall reimburse the Contractor for any cost the Contractor has incurred to the extent that the Contractor demonstrates that any defects were not caused by -

(a) the Contractor's neglect or default, or the neglect or default of any agent or subcontractor of his; or

(b) by any circumstances within his or their control.

(3) If the Contractor fails to comply with this Condition, the Employer may do anything necessary to make good any defects notified to the Contractor. All the costs and expenses reasonably incurred by the Employer in doing so shall be recoverable from the Contractor.

(4) In the case of any defects which have been made good under this Condition, the relevant Maintenance Period shall apply to the remedial works in full from the date of making good.

22
Occupier's rules and regulations (only applicable if stated in Abstract of Particulars)

The Contractor shall comply with the occupier's rules and regulations which have been provided to him or made available to him for inspection, both in respect of the Site and in respect of any larger premises of which the Site forms part. The Contractor shall comply with any changes to those rules and regulations notified to him as an Instruction under Condition 40 (PM's Instructions) during the execution of the Works.

23
Discrimination

(1) The Contractor shall not unlawfully discriminate within the meaning and scope of the provisions of the Race Relations Act 1976, the Sex Discrimination Acts 1975 and 1986 or the Sex Discrimination (Northern Ireland) Orders 1976 and 1988; and shall, if the Works are in Northern Ireland, conform with the requirements of the Fair Employment (Northern Ireland) Acts 1976 and 1989.

(2) The Contractor shall take all reasonable steps to ensure the observance of the provisions of paragraph (1) by all his employees, agents and subcontractors.

24
Corruption

(1) The Contractor shall not by himself, or in conjunction with any other person -

(a) corruptly solicit, receive or agree to receive, for himself or for any other person; or

(b) offer or agree to give to any person in the Employer's service, or any consultant or contractor who has a contract with the Employer;

any gift or consideration of any kind as an inducement or reward for doing or not doing anything, or for showing favour or disfavour to any person, in relation to this Contract or any other contract to which the Employer is a party.

(2) The Contractor shall not enter into this or any other contract with the Employer in connection with which commission has been paid or agreed to be paid by him or on his behalf or to his knowledge unless, before any such contract is made, particulars of any such commission, and of the terms and conditions of any agreement for the payment thereof, have been disclosed in writing to the Employer.

(3) The Employer may by notice determine the Contract if -

(a) he is reasonably satisfied that the Contractor or anyone employed by him or acting on his behalf (whether with or without the knowledge of the Contractor) is in breach of this Condition; or

(b) the Contractor or anyone employed by him or acting on his behalf is convicted of any offence under the Prevention of Corruption Acts 1889 to 1916 in relation to this Contract or any other contract to which the Employer is a party.

(4) If the Employer so determines the Contract, then (without prejudice to any powers conferred by Condition 51 (Recovery of sums) the Employer shall be entitled to recover from the Contractor the amount or value of any such gift, consideration or commission.

(5) In this Condition, all references to the Employer or the Contractor shall be deemed to include a reference to each member of their respective Groups.

25
Records

(1) The Contractor shall for the purposes of the Contract keep such records as may be reasonably necessary for the QS, the PM or the Employer to ascertain or verify any claims made or to be made by the Contractor or any sums to be paid to the Contractor under or in connection with the Contract.

(2) In order that the PM and QS may discharge their respective functions under the Contract, the Contractor shall afford them access to the records mentioned in paragraph (1) and supply them with the information (including means to interpret the records) they may require.

SECURITY

26 Site admittance

(1) The Contractor shall take the steps reasonably required by the PM to prevent unauthorised persons being admitted to the Site. If the PM gives the Contractor notice that any person is not to be admitted to the Site, the Contractor shall take all reasonable steps to prevent that person being admitted.

(2) If and when Instructed by the PM, the Contractor shall give to the PM a list of names and addresses of all persons who are or may be at any time concerned with the Works or any part of them, specifying the capacities in which they are so concerned, and giving such other particulars as the PM may reasonably require.

(3) The Contractor shall bear the cost of any notice, Instruction or decision of the PM under this Condition.

27 Passes (only applicable if stated in Abstract of Particulars)

Where passes are required for admission to the Site, the PM shall issue them to the Contractor. The Contractor shall submit to the Employer for his approval a list of the names of the workpeople and any other information the Employer reasonably requires for this purpose. The passes shall be returned at any time on the demand of the Employer and in any case on the completion of the Works.

28 Photographs (only applicable if stated in Abstract of Particulars)

The Contractor shall not at any time take any photograph of the Site or the Works or any part of them, and shall take all reasonable steps to ensure that no such photographs shall at any time be taken or published or otherwise circulated by any employee, agent or subcontractor of his, unless the Contractor has obtained the prior written consent of the PM.

29 Official secrets and confidentiality

(1) The Contractor shall take all reasonable steps to ensure that all persons employed by him or his subcontractors in connection with the Contract are aware of the Official Secrets Act 1989 and, where appropriate, of the provisions of Section 11 of the Atomic Energy Act 1946, and that these Acts apply to them during the execution of the Works and after the completion of the Works or earlier determination of the Contract.

(2) Any information concerning the Contract obtained either by the Contractor or by any person employed by him in connection with the Contract is confidential and shall not be used or disclosed by the Contractor or by any such person except for the purposes of the Contract.

MATERIALS AND WORKMANSHIP

30
Vesting

(1) The Works and any Things on the Site in connection with the Contract, the ownership of which the Contractor is able to transfer, or which vest in him under any contract, shall become the property of and vest in the Employer.

(2) Subject to Conditions 19 (Loss or damage) and 65 (Other works), the Employer shall not be responsible or chargeable for any Thing lost, stolen, damaged, destroyed or removed from the Site, or which is in any way unfit or unsuitable for its purpose.

(3) The Contractor shall be responsible for the protection and preservation of the Works and any Things brought on the Site until the completion of the Works or the determination of the Contract.

(4) No Things shall be removed from the Site before completion of the Works without the written consent of the PM. The PM may Instruct or permit the Contractor in writing at any time to remove from the Site any Things which are unused or which have been rejected by the PM, and the Contractor shall at his own expense forthwith remove them. Once so removed from the Site, the Things shall re-vest in the Contractor.

(5) In this Condition, all references to the Employer or the Contractor shall be deemed to include a reference to each member of their respective Groups.

31
Quality

(1) The Contractor shall execute the Works in accordance with the Contract and -

   (a) with diligence;

   (b) in accordance with the Design;

   (c) in accordance with the Programme;

   (d) with all reasonable skill and care;

   (e) in a workmanlike manner; and

   (f) in accordance with any quality control or assurance system, standards, and/or procedures specified in the Employer's Requirements.

(2) The Contractor warrants that all Things for incorporation, with the sole exception of Things for incorporation chosen or selected by the Employer by means of a statement by or on behalf of the Employer in the Contract or in a VI, shall be fit for their intended purposes, and shall conform to the requirements and standards specified in the Employer's Requirements or, where no such standards are specified, to the appropriate standards and/or standard codes of practice.

(3) The Contractor shall provide such samples as are specified in the Employer's Requirements, or as are reasonably required by the PM, to enable the PM to satisfy himself that the Contractor is complying with his obligations under the Contract. Such samples are to be provided in sufficient time so as to allow a reasonable period for approval before the carrying out of the relevant work. The Contractor shall thereafter ensure that all relevant Things for incorporation and/or standards of workmanship are of a standard no lower than the approved sample. No such approval shall relieve the Contractor of any liability.

(4) The Contractor shall notify the PM before incorporation of any Things that the Contractor considers should not be incorporated.

(5) The Contractor shall when requested by the PM demonstrate that he is performing his duties under paragraphs (1) and (2). The PM shall have power at any time to inspect and examine any part of the Works or inspect, examine and test any Things for incorporation either on the Site, or at any factory or workshop or other place where any such Thing is being constructed or manufactured, or at any place where it is lying, or from which it is being obtained. The Contractor shall give the PM the assistance and facilities he may reasonably require for any such inspection and examination. The PM may reject any Thing for incorporation which does not conform with the Employer's Requirements, or which, even though conforming, is not of good quality or is not fit for its purpose.

(6) The PM may arrange for an independent expert to test whether any Thing for incorporation is fit for use in the Works, of good quality, and conforms to the requirements of the Contract. The reasonable costs incurred by the Employer in arranging for an expert to carry out any test shall be borne by the Contractor, if the test results disclose that the Thing tested does not conform with the provisions of the Contract. The Contractor shall also bear the reasonable cost of any further tests reasonably required to monitor quality, following negative test results.

(7) The Contractor shall, at his own cost, replace, rectify or reconstruct -

(a) the Works, or any part of them, which do not conform with the Contract; and

(b) any Things for incorporation which do not conform with the Contract, and which have been rejected by the PM.

(8) If the Contract requires that the Contractor shall be accredited in accordance with any quality control or assurance scheme or system, the Contractor warrants that he is so accredited, and undertakes that he will continue to be so accredited until after the PM has issued a certificate under Condition 39 (Certifying completion) when the Contractor has complied with Condition 21 (Defects in Maintenance Periods). If the Contractor ceases for any reason to be so accredited during that period, he shall by notice immediately inform the PM, giving full particulars of the circumstances, and of his proposals to reinstate the relevant accreditation, or to obtain an alternative accreditation.

32
Excavations

(1) Except as otherwise provided by the Contract, material and objects of any kind obtained from work on the Site (including, without limitation, from excavations, demolition or dismantling) shall remain or become the property of the Employer.

(2) When the Employer's property is permitted to be used in substitution for any Things (whether or not for incorporation), which the Contractor would otherwise have provided, the QS shall ascertain the amount of any saving in the cost of the execution of the Works. The Contract Sum shall be reduced by the amount of any saving.

(3) All objects which are, or appear to be, fossils, antiquities, or likely to have interest or value, found on the Site or in carrying out excavations in the execution of the Works, shall remain or become the property of the Employer. Upon the discovery of any such object, the Contractor shall forthwith -

(a) take all practicable measures not to disturb the object;

(b) cease work, if the continuance of work would endanger, or disturb, the object, or prevent or impede its excavation or removal;

(c) take all necessary steps to preserve the object in the exact position and condition in which it was found; and

(d) inform the PM of the discovery and precise location of the object.

(4) Any Instructions issued by the PM in relation to any object mentioned in paragraph (3), may require the Contractor to permit the examination, excavation or removal of the object by a third party.

COMMENCEMENT, PROGRAMME, DELAYS AND COMPLETION

33

Programme

(1) The Contractor warrants that the Programme shows the Design Process Event and the time for its certification, the sequence in which the Contractor proposes to execute the Works, details of any temporary work, method of work, labour and plant proposed to be employed, and events, which, in his opinion, are critical to the satisfactory completion of the Works; that the Programme is achievable, conforms with the requirements of the Contract, permits effective monitoring of progress, and allows reasonable periods of time for the provision of information required from the Employer; and that the Programme is based on a period for the execution of the Works to the Date or Dates for Completion.

(2) Subject to Conditions 35 (Progress meetings), 37 (Early possession) and 38 (Acceleration), the Contractor may at any time submit for the PM's agreement proposals for the amendment of the Programme. The agreement of the PM to any proposal for the amendment of the Programme shall not relieve the Contractor of any liability which he has under the Contract. In particular, without limitation, the submission by the Contractor of any proposal for the amendment of the Programme showing a period for the execution of the Works extending beyond the Date or any of the Dates for Completion shall not constitute a notice from the Contractor requesting an extension of time for the completion of the Works or of any Section; and the agreement of the PM to any such amendment shall not constitute, or be evidence of, or in support of, any extension of time for the completion of the Works or of any Section.

34

Commencement and completion

(1) Within the period specified in the Abstract of Particulars, the Employer shall have an absolute discretion whether or not to proceed with the Contract. Within that period, the Employer shall either notify the Contractor that the Employer has decided not to proceed with the Contract, or notify the Contractor when he may take possession of the Site, or those parts of the Site defined in the Contract. Any such notification to the Contractor when he may take possession of the Site, or those parts of the Site defined in the Contract, shall constitute acceptance by the Employer of any then existing or future determination of the Contract Sum by the QS in accordance with Condition 10B (Design process and Contract Sum). Any date so notified shall be not more than 14 Days after the end of the period specified in the Abstract of Particulars. The Date or Dates for Completion shall be calculated from the date so notified. The Contractor shall accordingly take possession of the Site, or such parts of the Site defined in the Contract. The Contractor shall, subject to Condition 11 (Statutory notices and CDM Regulations), proceed with diligence and in accordance with the Programme or as may be Instructed by the PM, so that the whole of the Works or any relevant Section shall be completed in accordance with the Contract by the Date or Dates for Completion.

(2) The Contractor shall, at all times, keep the Site tidy and free from débris, litter, and rubbish and shall, not later than the completion of the Works, remove from the Site all Things for incorporation in the Works or any relevant Section which are unused, together with all Things not for incorporation. The Contractor shall, by the due date, clear and remove all rubbish and deliver up the Site and the Works in all respects to the satisfaction of the PM. The Contractor shall comply at his own cost with any Instructions relating to the removal of any Things and rubbish.

34A

Deemed determination

(1) If the Employer, within the period specified in the Abstract of Particulars, and pursuant to Condition 34 (Commencement and completion), either notifies the Contractor that the Employer had decided

not to proceed with the Contract, or does not notify the Contractor when he may take possession of the Site, or those parts of the Site defined in the Contract, the Contract shall thereby be deemed determined.

(2) In that event, the Contractor shall -

(a) cease forthwith to carry out any work whatsoever in connection with Contract and shall deliver up to the Employer all documents and particulars comprising the Design or parts of the Design; and

(b) be entitled to be paid in accordance with the Contract any balance of the Design Fee, or of that part of the Design Fee which relates to work he has executed, and any balance in respect of the execution of any other part of the Works Instructed under Condition 10B (Design process and Contract Sum).

Payment in accordance with this paragraph shall be the Contractor's sole and exclusive right or remedy in respect of the deemed determination of the Contract under this Condition.

35 Progress meetings

(1) The Contractor's agent shall attend regular progress meetings to assess the progress of the Works or any relevant Section, and to facilitate their due and satisfactory completion by the Date or Dates for Completion.

(2) An initial progress meeting shall be held, and thereafter progress meetings shall be held each month, unless Instructed to the contrary. The PM shall specify the time and place of progress meetings.

(3) The Contractor shall submit to the PM, 5 Days before each progress meeting, a written report which shall -

(a) describe the progress of the Works by reference to the Programme and relevant Instructions;

(b) specify all outstanding requests by the Contractor for drawings, levels or other information;

(c) explain any new circumstances, arising since any previous progress meeting, which in his opinion have delayed, or may delay, completion of the Works or a Section, or may increase the cost to the Employer of the Works or a Section, estimating any increase in such cost;

(d) refer to any request for an extension of time under Condition 36 (Extensions of time) since the previous progress meeting; and

(e) set out any re-programming proposals to ensure that completion of the Works or any Section will be achieved by the relevant Date for Completion.

(4) The PM shall, within 7 Days after each progress meeting, give the Contractor a written statement which specifies -

(a) by reference to the Programme the extent to which he considers the Works are on time, delayed or early;

(b) the matters which the PM considers have delayed, or are likely to delay, due completion of the Works or Section, or may increase the cost to the Employer of the Works or a Section, estimating any increase in such cost;

(c) the steps which the PM has agreed with the Contractor to reduce or eliminate the effects of any such delay, or to reduce to eliminate any estimated increase in the cost to the Employer of the Works or a Section;

(d) the situation in respect of applications for, and awards of, extensions of time under Condition 36 (Extensions of time); and

(e) his response to outstanding requests for drawings, levels or other information.

36 Extensions of time

(1) Where the PM receives notice requesting an extension of time from the Contractor (which shall include the grounds for his request), or where the PM considers that there has been or is likely to be a delay which will prevent or has prevented completion of the Works or any Section by the relevant Date for Completion (in this Condition called 'delay'), he shall, as soon as possible and in any event within 42 Days from the date any such notice is received, notify the Contractor of his decision regarding an extension of time for completion of the Works or relevant Section.

(2) The PM shall award an extension of time under paragraph (1) only if he is satisfied that the delay, or likely delay, is or will be due to -

(a) a change in the Employer's Requirements;

(b) any act, neglect or default of the Employer, the PM or any other person for whom the Employer is responsible (not arising because of any default or neglect by the Contractor or by any employee, agent or subcontractor of his);

(c) any strike or industrial action which prevents or delays the execution of the Works, and which is outside the control of the Contractor or any of his subcontractors;

(d) an Accepted Risk or Unforeseeable Ground Conditions;

(e) any other circumstances (not arising because of any default or neglect by the Contractor or by any employee, agent or subcontractor of his, and other than weather conditions), which are outside the control of the Contractor or any of his subcontractors, and which could not have been reasonably contemplated under the Contract;

(f) failure of the Planning Supervisor to carry out his duties under the CDM Regulations properly; or

(g) the exercise by the Contractor of his rights under Condition 52 (Suspension for non-payment).

(3) The PM shall indicate whether his decision is interim or final. The PM shall keep all interim decisions under review until he is satisfied from the information available to him that he can give a final decision.

(4) No requests for extensions of time may be submitted after completion of the Works. The PM shall in any event come to a final decision on all outstanding and interim extensions of time within 42 Days after completion of the Works. The PM shall not be entitled in a final decision to withdraw or reduce any interim extension of time already awarded, except to take account of any authorised omission from the Works or any relevant Section that he has not already allowed for in an interim decision.

(5) The Contractor may within 14 Days from receipt of a decision, submit a claim to the PM specifying the grounds which in his view entitle him to an extension or further extension of time. The PM shall by notice give his decision on a claim within 28 Days of its receipt.

(6) The Contractor must endeavour to prevent delays and to minimise unavoidable delays, and to do all that may be required to proceed with the Works. The Contractor shall not be entitled to an extension of time where the delay or likely delay is, or would be, attributable to the negligence, default, improper conduct or lack of endeavour of the Contractor.

37

Early possession

(1) The Employer shall be entitled, before the completion of the Works, to take possession of any part of the Works (in this Condition referred to as a 'completed part') which is certified by the PM as having been completed in accordance with the Contract and is either -

(a) a Section; or

(b) any other part of the Works in respect of which the parties agree, or the PM has given an Instruction, that possession shall be given before the completion of the Works or the relevant Section;

and the completed part, on and after the date on which the certificate is given, shall no longer form part of the Works for the purposes of Conditions 19 (Loss or damage) and 30 (Vesting).

(2) The provisions of Condition 21 (Defects in Maintenance Periods) shall have effect in relation to a completed part as if the Maintenance Period or Periods in respect of the completed part, or any subcontract works comprised in it, commenced on the date of certification under paragraph (1).

(3) As soon as possible after certification under paragraph (1), the PM shall certify the value of the completed part and his estimates for the purposes of paragraphs (4) and (5).

(4) Condition 55 (Liquidated damages) shall have effect notwithstanding that the Employer has taken possession of a completed part. Where the completed part comprises part of the Works as mentioned in subparagraph (1)(b), the rate of liquidated damages specified in the Abstract of Particulars in respect of the Works or the relevant Section shall be reduced by the same ratio as the value of the completed part bears to what the value of the Works or of the relevant Section will be when completed in accordance with the Contract, as estimated by the PM.

(5) The retention accumulated in accordance with Condition 48 (Advances on account) shall be apportioned by the PM with effect from the date of the certificate given under paragraph (1), so that the share of the retention apportioned in respect of a completed part shall bear the same ratio to the whole of the retention as the value of the completed part bears to the value of the Works which at that date have been completed in accordance with the Contract, as estimated by the PM.

(6) The Employer shall pay to the Contractor -

(a) one half of the share apportioned in accordance with paragraph (5) in respect of the completed part; and

(b) the remaining one half of that share, when the PM has certified after the appropriate Maintenance Period that in respect of the completed part the Contractor has complied with Condition 21 (Defects in Maintenance Periods).

38

Acceleration

(1) If the Employer wishes to achieve completion of the Works or any Section before the Date or Dates for Completion, he shall direct the Contractor to submit to him within the period specified in the direction -

(a) the Contractor's priced proposals for achieving the accelerated completion date, together with any consequential amendments to the Programme; or

(b) the Contractor's explanation why he is unable to achieve the accelerated completion date.

(2) If the Employer accepts the Contractor's proposals he shall specify -

(a) the accelerated Date for Completion of the Works or any relevant Section:

(b) the amendments to the Programme, including any relevant critical paths and any supporting documentation;

(c) the amount by which the Contract Sum shall be adjusted;

(d) a revised Milestone or Stage Payment Chart or Charts, if necessary; and

(e) any other relevant amendment to the Contract which has been agreed with the Contractor.

(3) The Contractor may at any time submit to the Employer proposals for completing the Works or any Section before the Date or Dates for Completion. The Employer undertakes to consider any such proposals and if he accepts them to take action as stated in paragraph (2).

38A

Bonuses (only applicable if stated in Abstract of Particulars)

If a rate of bonus for early completion of the Works or a Section has been specified in the Abstract of Particulars, and the Works or the relevant Section are or is completed before the relevant Date for Completion, the Employer shall pay the Contractor bonus at the rate specified in the Abstract of Particulars for each Day falling between the date certified by the PM under Condition 39 (Certifying completion) as the date of completion of the Works, or of the relevant Section, and the relevant Date for Completion.

39

Certifying completion

The PM shall certify the date when the Works, or any Section, or any completed part within the meaning of Condition 37 (Early possession), are completed in accordance with the Contract. Such completion shall include sufficient compliance by the Contractor with Condition 11(7) (Statutory notices and CDM Regulations). After the end of the last Maintenance Period to expire, he shall issue a certificate when the Contractor has complied with Condition 21 (Defects in Maintenance Periods).

INSTRUCTIONS AND PAYMENT

40

PM's Instructions

(1) The PM may from time to time issue further drawings, details, instructions, directions and explanations, all or any of which shall be treated for the purposes of the Contract as Instructions, including Variation Instructions. The Contractor shall comply forthwith with any Instruction.

(2) Instructions may be given in relation to all or any of the following matters -

(a) a change in the Employer's Requirements;

(b) the resolution of such discrepancies as are referred to in Condition 2 (Contract documents);

(c) the removal from the Site of any Things for incorporation and their substitution with any other Things;

(d) the removal and/or re-execution of any work executed by the Contractor;

(e) the order of execution of the Works or any part of them;

(f) the hours of working and the extent of overtime or night work to be adopted;

(g) the suspension of the execution of the Works or any part of them;

(h) the replacement of any person employed in connection with the Contract;

(i) the opening up for inspection of any work covered up;

(j) the amending and making good of any defects under Condition 21 (Defects in Maintenance Periods);

(k) (not used)

(l) the execution of any emergency work as mentioned in Condition 54 (Emergency work);

(m) the use or disposal of material obtained from excavations, demolition or dismantling on the Site;

(n) the actions to be taken following discovery of fossils, antiquities or objects of interest or value;

(o) measures to avoid nuisance or pollution;

(p) quality control accreditation of the Contractor as mentioned in Condition 31 (Quality); and

(q) any other matter which the PM considers necessary or expedient.

(3) All Instructions shall be in writing except those under subparagraphs (b), (d), (g) and (l) of paragraph (2), which may be given orally. Oral Instructions shall be immediately effective in accordance with their terms, but shall be confirmed in writing by the PM within 7 Days. Written Instructions shall be given to the Contractor's agent. Oral Instructions may be given to such employee or agent of the Contractor as the PM thinks fit, but shall be confirmed in writing to the Contractor's agent. The Contractor's agent shall immediately acknowledge receipt of any written Instruction, and of every written confirmation of an oral Instruction.

(4) (Not used)

(5) The PM may include in a VI a requirement for the Contractor to submit to the QS not later than 21 Days from the receipt of that Instruction a written quotation of the lump sum total price of

complying with it. The PM may make any VI conditional upon agreement of such a lump sum price, pending which agreement the Contractor is not to begin complying with the VI.

41

Valuation of Instructions - Principles

(1) The value (if any) of any VI shall be determined in accordance with Condition 42 (Valuation of Variation Instructions) and the value (if any) of any other Instruction in accordance with Condition 43 (Valuation of other Instructions).

(2) The value of any Instruction shall include any disruption to or prolongation of both varied and unvaried work.

(3) The value of any Instruction shall be added to, or as the case may be deducted from, the Contract Sum, except where the Instruction was necessary because of any default or neglect by the Contractor or by any employee, agent or subcontractor of his.

(4) The Contractor shall supply the QS with any information required by the QS to enable him to value a VI or determine the expense (if any) of complying with any other Instruction.

42

Valuation of Variation Instructions

(1) The value of a VI shall be determined -

- (a) by acceptance by the PM of a lump sum quotation prepared by the Contractor and submitted to the QS in accordance with Condition 40(5) (PM's Instructions) or such other lump sum as may be agreed by negotiation on such a quotation; or
- (b) by valuation by the QS in accordance with paragraphs (5) to (11).

(2) Any lump sum quotation submitted by the Contractor to the QS in accordance with Condition 40(5) (PM's Instructions) shall indicate how the lump sum was calculated by showing separately the amounts attributable to -

- (a) complying with that Instruction; and
- (b) any disruption to or prolongation of varied and unvaried work consequential upon compliance with the VI.

The Contractor shall include with his quotation such other information as will enable the QS to evaluate that quotation.

(3) The PM shall notify the Contractor, not later than 21 Days from the receipt of any such lump sum quotation, whether or not it is accepted or, if it is not acceptable, whether he is prepared to agree any other lump sum. If accepted, the aggregate amount specified in that quotation, or otherwise agreed between the Employer and the Contractor, shall be the full sum to which the Contractor is entitled for complying with that VI.

(4) Should -

- (a) the Contractor fail to provide a lump sum quotation in accordance with Condition 40(5) (PM's Instructions); or
- (b) the Employer and the Contractor fail to agree by negotiation, and no agreement for independent assessment is reached;

then the PM shall instruct the QS to value the VI.

(5) Where the QS is required to value a VI he shall do so -

(a) by measurement and valuation at the rates and prices in the Pricing Document for similar work; or

(b) if it is not possible to value as in subparagraph (a), then by measurement and valuation at rates and prices deduced or extrapolated from the rates and prices in the Pricing Document; or

(c) if it is not possible to value as in subparagraph (b), then by measurement and valuation at fair rates and prices, having regard to current market prices; or

(d) if it is not possible to value by any of the preceding methods of measurement and valuation, then by the value of the materials used and plant and labour employed in accordance with the basis of charge for daywork described in the Contract.

(6) If, in the opinion of the QS, the VI prolongs and/or disrupts work not within the direct scope of the VI, then the QS shall adjust to the rates for such work as he considers appropriate.

(7) The Contractor shall submit the information mentioned in Condition 41(4) (Valuation of Instructions - Principles) not later than 14 Days after its being requested by the QS.

(8) The QS shall, not later than 28 Days from the receipt of the information mentioned in paragraph (7), notify the Contractor of his valuation of the VI.

(9) If the Contractor disagrees with the whole or part of the QS's valuation he shall, within 14 Days of the QS's notification under paragraph (8), give his reasons for disagreement and his own valuation. In any other case he shall be treated as having accepted the notification under paragraph (8), and no further claim shall be made by him in respect of the VI.

(10) Any percentage or lump sum adjustments made in the pricing of the Pricing Document shall be deemed to be applicable to the pricing of all relevant VI's.

(11) Where an alteration in, or addition to, the Works would otherwise fall to be valued using rates and prices in the Pricing Document in accordance with subparagraphs (5)(a) or (b), but the QS is of the opinion that the relevant VI was issued at such a time or was of such content as to make it unreasonable for the alteration or addition to be so valued, he shall ascertain the value by measurement and valuation at fair rates and prices, having regard to current market prices.

(12) Where the QS has determined that work shall be valued in accordance with subparagraph (5)(d), the Contractor shall give reasonable notice to the QS before commencing the work, and shall deliver to him, by the end of the week following that in which the work was done, vouchers in the form required by the QS, which specify separately the labour, materials and plant for that week.

(13) Where, as a result of a VI, the Contractor makes a saving in the cost of executing the Works, the Contract Sum shall be decreased by the amount of the saving as determined by the QS.

43

Valuation of other Instructions

(1) Where as the result of an Instruction, not being a VI, the Contractor -

(a) properly and directly incurs any expense beyond that provided for in, or reasonably contemplated by, the Contract; or

(b) makes any saving in the cost of executing the Works;

the Contract Sum shall, subject to Condition 41(3) (Valuation of Instructions - Principles), be increased by the amount of the expense, or decreased by the amount of the saving, in either case as determined by the QS.

(2) The Contractor shall submit the information mentioned in Condition 41(4) (Valuation of Instructions - Principles) within 28 Days of complying with the relevant Instruction. Within 28 Days of receiving such information, the QS shall notify the Contractor of the amount he has determined.

(3) If the Contractor disagrees with the amount notified under paragraph (2), he shall within 14 Days of receipt notify the QS of his reasons for disagreement and of his own estimate of the correct amount. If the Contractor does not so notify, he shall be regarded as having accepted the amount determined by the QS, and may make no further claim in respect of the Instruction.

(4) In this Condition 'expense' shall mean money expended by the Contractor, but shall not include any sum expended, or loss incurred, by him by way of interest or finance charges however described.

44

(Not used)

45

VAT

(1) All sums payable by or to the Employer or the Contractor are exclusive of Value Added Tax ('VAT'). Where VAT is chargeable on such sums, the payer shall pay, upon production of a valid VAT invoice by the payee, such VAT in addition to such sums.

(2) If the Contractor fails to carry out his obligations under the Contract and the Employer employs some other contractor to fulfil them, and a payment in respect of VAT is made or falls to be made to that other contractor, then the Employer shall be entitled to recover from the Contractor any VAT (which he is not otherwise able to recover) additional to what he would have paid had the Contractor carried out his obligations under the Contract.

(3) Where a party is liable to reimburse or indemnify the other party for costs incurred by that other party, the amount to be paid shall not include any VAT charged on such costs, save where the payee is unable to recover such VAT from HM Customs & Excise as input tax.

46

Prolongation and disruption

(1) If the Contractor properly and directly incurs any expense which he would not otherwise have incurred by reason of -

- (a) the execution of works pursuant to Condition 65 (Other works); or
- (b) any delay in being given possession of the Site or part of it, or any delay in respect of any of the matters specified in paragraph (2); or
- (c) any advice from the Planning Supervisor other than that properly required by the CDM Regulations;

which unavoidably results in the regular progress of the Works or of any part of them being materially disrupted or prolonged and which is beyond that provided for or reasonably contemplated by the Contract, and is not a consequence of any default or neglect on the part of the Contractor, or by any employee, agent or subcontractor of his, the Contract Sum shall be increased by the amount of that expense as determined by the QS.

(2) The matters referred to in subparagraph (1)(b) are -

- (a) any decisions, confirmations, consents, drawings, schedules, levels or other design information to be provided by the PM;
- (b) the execution of any work or the supply of any Thing by the Employer or ordered from somebody other than the Contractor; and

(c) any direction or Instruction from the Employer or the PM regarding the issue of any pass to any person (if required under Condition 27 (Passes)), provided that the Employer and the PM shall be entitled to a reasonable time for consideration and decision in respect of those matters, and this Condition shall not fetter their proper discretion under that Condition.

(3) The Contract Sum shall not be increased under paragraph (1) unless -

(a) the Contractor, immediately upon becoming aware that the regular progress of the Works or any part of them has been or is likely to be disrupted or prolonged, has given notice to the PM specifying the circumstances causing or expected to cause that disruption or prolongation and stating that he is, or expects to be, entitled to an increase in the Contract Sum under that paragraph; and

(b) the Contractor, as soon as reasonably practicable, and in any case within 56 Days of incurring the expense, provides to the QS full details of all expenses incurred and evidence that the expenses directly result from the occurrence of one of the events described in paragraph (1).

(4) Subject to paragraph (3), the Contract Sum shall be increased in accordance with subparagraph (1)(b) only where the Employer has failed to supply an item or act -

(a) by a date agreed beforehand with the Contractor; or

(b) within any reasonable period specified in a notice given by the Contractor to the Employer or the PM for the supply of the item or taking the action.

(5) The QS shall, not later than 28 Days from the receipt of the information referred to in subparagraph (3)(b), notify the Contractor of his decision under this Condition.

(6) In this Condition 'expense' shall mean money expended by the Contractor, but shall not include any sum expended, or loss incurred, by him by way of interest or finance charges however described.

47
Finance charges

(1) The Employer shall pay the Contractor an amount by way of interest or finance charges (hereafter together called 'finance charges') only in the event that money is withheld from him under the Contract because either -

(a) the Employer, PM or QS has failed to comply with any time limit specified in the Contract or, where the parties agree at any time to vary any such time limit, that time limit as varied; or

(b) the QS varies any decision of his which he has previously notified to the Contractor.

(2) Finance charges shall be calculated as a percentage of the amounts which would have been paid to the Contractor if any of the events mentioned in paragraph (1) had not occurred. The rate at which finance charges shall be payable shall be the percentage stated in the Abstract of Particulars over the rate charged during the relevant period by the Bank of England for lending money to the clearing banks, compounded with effect from 31 March, 30 June, 30 September and 31 December.

(3) Finance charges shall be payable in respect of any period commencing with the date on which, but for a failure or variation mentioned in paragraph (1), money properly due under the Contract should have been certified, and ending with the date on which it was certified for payment under Condition 50 (Certifying payments).

(4) When calculating finance charges the QS shall take into account any overpayment made to the Contractor as a result of circumstances described in subparagraph (1)(b).

(5) The Employer shall not be liable to pay any finance charges which result from -

(a) any act, neglect or default of the Contractor or any of his subcontractors;

(b) any failure by the Contractor or any of his subcontractors to supply the PM or the QS with any relevant information; or

(c) any disagreement about the Final Account.

(6) The Employer and the Contractor agree that, when they entered into the Contract, neither of them had knowledge of any special facts or circumstances which would entitle the Contractor to be paid interest or finance charges, except in the circumstances mentioned in paragraph (1).

(7) The respective powers of the adjudicator and arbitrator to award interest shall be in addition to, and not in derogation from, this Condition.

48

Advances on account

(1) The Contractor shall following certification under Condition 50 (Certifying payments) be entitled to be paid advances during the execution of the Works.

(2) The amount of each advance to be certified shall, subject to Condition 48A (Retention payment bond), be the total of the following sums -

Alternative A (Stage Payment Chart)

(a) (i) In respect of the Design Fee, and the execution of any other part of the Works Instructed under Condition 10B (Design process and Contract Sum), 100% of the value of the work executed. The Contractor shall submit applications in writing to the PM for the payment of such advances on account, supported by a valuation of the relevant work, not later than 7 Days before such certificate is due under Condition 50 (Certifying payments). Unless and until the Contractor timely submits such an application and valuation, he shall not be entitled to certification or payment in respect of the relevant work.

(ii) In respect of the balance of the Contract Sum, 95% of the proportion of the sum specified as the proportion of the Contract Sum payable for the relevant month according to the Stage Payment Chart or Charts.

Alternative B (Milestone Payment Chart)

(a) (i) In respect of the Design Fee, and the execution of any other part of the Works Instructed under Condition 10B (Design process and Contract Sum), 100% of the value of the work executed. The Contractor shall submit applications in writing to the PM for the payment of such advances on account, supported by a valuation of the relevant work, not later than 7 Days before such certificate is due under Condition 50 (Certifying payments). Unless and until the Contractor timely submits such an application and valuation, he shall not be entitled to certification or payment in respect of the relevant work.

(ii) In respect of the balance of the Contract Sum, unless the Milestone Payment Chart provides otherwise, 95% of the cumulative value (not previously paid) of the Works according to the Milestone Payment Chart at the relevant Milestone achieved. Unless the Milestone Payment Chart provides otherwise, and specifies what payment is to be made upon the achievement of each Milestone, the Contractor shall not be entitled to any payment upon achievement of a Milestone until he has also achieved all the preceding Milestones set out in the Milestone Payment Chart. The Contractor shall submit applications in writing to the PM for payment under this subparagraph, upon the achievement of each Milestone. Unless and until the Contractor submits such an application, he shall not be entitled to certification or payment in respect of the relevant sum. If the Abstract of Particulars so specifies, such an application shall also include all amounts due under subparagraphs (2)(b)-(h), and the Contractor shall not be entitled to certification or payment of an advance in respect of such amounts except as included in such application. Paragraph (3) shall not apply.

Alternative C (Valuation)

(a) (i) In respect of the Design Fee, and the execution of any other part of the Works Instructed under Condition 10B (Design process and Contract Sum), 100% of the value of the work executed, and in respect of the balance of the Contract Sum, 95% of the value of the work executed on the Site (other than such work referred to in subparagraphs (c) and/or (d)); and 95% of the value of any Things for incorporation which have been reasonably delivered to the Site and are adequately stored and protected against damage by weather and other causes, but which have not been incorporated in the Works.

(ii) Where any Things for incorporation on account of which an advance has been made under this subparagraph are incorporated in the Works, the amount of such advance may be deducted from the next or any subsequent payment made under the Contract.

(iii) The Contractor shall submit applications in writing to the PM for payment of advances on account of work executed and Things for incorporation which have been delivered to the Site, supported by a valuation of such work and Things, not later than 7 Days before such certificate is due under Condition 50 (Certifying payments). Unless and until the Contractor timely submits such an application and valuation, he shall not be entitled to certification or payment of an advance in respect of the relevant work or Things. Paragraph (3) shall not apply.

(b) 100% of any amount calculated under Condition 38A (Bonuses) (if applicable).

(c) 100% of any amount agreed under Condition 42(1)(a) (Valuation of Variation Instructions) in respect of work completed in the relevant month.

(d) 100% of the agreed value or, failing agreement, 95% of the QS's valuation, under Condition 42(5) (Valuation of Variation Instructions) and Condition 43 (Valuation of other Instructions) in respect of work completed in the relevant month.

(e) 100% of any amount determined by the QS under Condition 46 (Prolongation and disruption) in respect of the relevant month.

(f) 100% of any amount calculated under Condition 47 (Finance charges).

(g) 95% of any amount calculated under Condition 48C (Payment for Things off-Site) (if applicable).

(h) Less, in each case, any sum agreed to be credited by the Contractor for old materials.

(3) Where the PM has recorded in a statement after a progress meeting that the Works are in delay or are ahead of Programme he shall by reference to the Stage Payment Chart or Charts adjust the Contractor's entitlement to payments in accordance with paragraph (2)(a).

(4) Before the payment of any advance or the issue of the final certificate for payment the Contractor shall, if requested by the PM, demonstrate to him that any amount due to a subcontractor or supplier of Things for incorporation which is covered by any previous advance has been paid. In any case where the PM is not satisfied the Employer may withhold payment to the Contractor of the amount in question until the PM is satisfied.

(5) The Employer shall accumulate as a retention the balance of any sum withheld under paragraph (2) which is less than 100% and shall continue to hold the entire beneficial interest therein.

48A

Retention payment bond (only applicable if stated in Abstract of Particulars)

If the Abstract of Particulars provides for the Employer to pay the Contractor advances on account without deduction of retention, provided that the amount of retention so foregone by the Employer shall not exceed the amount stated in the Abstract of Particulars ('the Retention Payment'), the Contractor shall be entitled to be paid advances on account without deduction of retention, after fulfilling the following conditions precedent -

(a) delivery to the Employer of a retention payment bond in the form prescribed by the Contract from the surety or sureties named in the tender in the amount of the Retention Payment;

(b) such delivery to occur not before and within 28 Days after the Employer notifies the Contractor when he may take possession of the Site, or those parts of the Site defined in the Contract, pursuant to Condition 34 (Commencement and completion), in respect of which period time shall be of the essence.

48B

Mobilisation payment (only applicable if stated in Abstract of Particulars)

(1) If the Abstract of Particulars provides for the Employer to pay the Contractor a mobilisation payment, the Contractor shall be entitled to be paid that payment within 14 Days of fulfilling the following conditions precedent -

(a) delivery to the Employer of a mobilisation payment bond in the form prescribed by the Contract from the surety or sureties named in the tender in the amount of the mobilisation payment;

(b) such delivery to occur not before and within 28 Days after the Employer notifies the Contractor when he may take possession of the Site, or those parts of the Site defined in the Contract, pursuant to Condition 34 (Commencement and completion), in respect of which period time shall be of the essence.

(2) The Employer shall be entitled to recover the mobilisation payment by deduction (in addition to any retention) of the percentage stated in the Abstract of Particulars either -

(a) from the proportion of the sum specified as the proportion of the Contract Sum payable for the relevant month according to the Stage Payment Chart or Charts; or

(b) from the cumulative value of the Works according to the Milestone Payment Chart at the relevant Milestone achieved in the relevant month; or

(c) from the cumulative value of the work executed on Site and of Things for incorporation delivered to the Site;

whichever is applicable under Condition 48 (Advances on account).

(3) Any part of the mobilisation payment not previously recovered by the Employer shall be immediately repayable by the Contractor to the Employer upon either -

(a) the certification by the PM under Condition 39 (Certifying completion) of the completion of the Works or the last Section thereof in respect of which completion is certified; or

(b) the determination of the Contract for any reason whatsoever, including (without limitation) breach by the Employer;

whichever occurs first.

48C

Payment for Things off-Site (only applicable if stated in Abstract of Particulars, *and not applicable in Scotland*)

(1) Where this Condition applies compulsorily, as specified in the Abstract of Particulars, the PM shall, and where this Condition applies voluntarily, as so specified, the PM may in his complete discretion, include in his certificates under Condition 50 (Certifying payments) 95% of his estimate of the value at that time (ascertained on the basis of fair and reasonable prices) of Things to which this Condition applies, and which have been vested in the Employer in accordance with this Condition. For the purposes of this Condition, 'Things' shall mean any Things for incorporation which are manufactured, assembled or constructed off the Site and which the PM considers are in accordance with the Contract and substantially ready for incorporation in the Works. Any sum so certified and paid may, when the relevant Things have been delivered to the Site, or re-vest in the Contractor for any reason whatsoever, be deducted from any subsequent advance, without prejudice and in addition to any other rights and remedies of the Employer.

(2) In order to transfer the property in Things, the Contractor shall -

(a) ensure that the Things have been properly and securely set aside at the factory or workshop or other place where any such Things have been manufactured, assembled or constructed or at any place where they are lying or from which they are being obtained;

(b) ensure that the Things have been suitably marked or otherwise identified so as to show that their destination is the Site, that they are the property of the Employer and, where relevant, to whose order they are held; and

(c) provide to the PM documentary evidence that the property in the Things has vested unconditionally in the Contractor.

(3) Upon the PM confirming that the Contractor has complied with the requirements of paragraph (2), the Contractor shall transfer to the Employer the property in the Things. Upon the PM approving that transfer, the Things shall vest in and become the absolute property of the Employer. The Things shall thenceforth be in the possession of the Contractor for the sole purpose of the performance of the Contract and delivering the completed Things to the Site for inclusion in the Works, and shall not be within the ownership or disposition of the Contractor.

(4) Approval by the PM for the purposes of this Condition shall be without prejudice to the power of the PM to reject any Things in accordance with Condition 31 (Quality). In the event of any Things being rejected by the PM in accordance with Condition 31 (Quality), they shall re-vest in the Contractor.

(5) The Contractor shall be solely responsible for, and shall take all responsible and proper steps for protecting, preserving and securing any Things held off the Site.

(6) The Contractor shall comply with any statutory regulations (whether or not binding on the Crown) which govern the storage and use of all Things off the Site.

(7) The Contractor shall indemnify the Employer against all claims and proceedings made or brought against the Employer in respect of any loss and/or damage which arises out of, or is any way connected with, the manufacture, assembly, construction, storage or transportation of any Things. For the purpose of this Condition, loss or damage includes -

(a) loss or damage to property;

(b) personal injury to, or the sickness or death of, any person;

(c) loss or damage to any Things; and

(d) loss of profits or loss of use suffered because of any loss or damage.

(8) If the Contract is determined the Employer may give to the Contractor, before the expiration of 28 Days from the date on which such determination takes effect, one or more notices stating that the Employer elects that all or any Things which have not been delivered to the Site shall re-vest in the Contractor. The Things specified in the notice shall re-vest upon service of that notice.

(9) With regard to Things which the Employer does not elect shall re-vest in the Contractor under paragraph (8) -

(a) the Contractor shall hand over to the Employer the Things, and if he fails to do so, the Employer may enter any premises and remove the Things and recover the cost of doing so from the Contractor; and

(b) subject to the terms of the Contract (in particular, without limitation, Condition 57 (Consequences of determination by Employer), if applicable to the determination of the Contract) the Employer shall pay a fair and reasonable price for the Things which are handed over to him by the Contractor or otherwise come into his possession, but shall be given credit for any payment therefor previously made under this Condition or otherwise.

(10) Any payment made by the Employer in respect of any Things which re-vest in the Contractor under paragraphs (4) or (8) shall be a sum recoverable in accordance with Condition 51 (Recovery of sums).

(11) The Contractor shall incorporate provisions equivalent to those provided in this Condition in every subcontract in which provision is to be made for the manufacture, assembly or construction of Things off the Site.

49
Final Account

(1) Upon completion of the Works in accordance with the Contract the Employer shall as soon as reasonably possible pay to the Contractor an amount equal to the difference between -

(a) the amount estimated by the Employer to be the Final Sum less one half the retention accumulated under Condition 48(5) (Advances on account); and

(b) the total amount of advances paid under Condition 48 (Advances on account).

(2) Within 6 months of the certified completion of the Works or the last part thereof under Condition 39 (Certifying completion), the QS shall forward one copy of the draft final account to the Contractor. If the draft final account shows a Final Sum greater than the Final Sum estimated by the Employer under paragraph (1)(a), the Employer shall correct any payment made under paragraph (1) as soon as possible, in accordance with the Final Sum shown in the draft final account. The Contractor shall within 3 months of receipt of the draft final account notify his agreement or disagreement with the draft final account. The Contractor shall specify in any notice of disagreement his reasons for disagreement and his own valuation.

(3) If the Contractor does not give notice of disagreement or gives notice but fails to specify adequate grounds for disagreement he shall be deemed to have agreed the draft final account as the Final Account.

(4) If before the end of the Maintenance Period, or where there is more than one the end of the last Maintenance Period to expire, the Final Sum has been calculated and agreed or has been treated as

having been agreed under paragraph (3), or in default of agreement has been determined in accordance with the Contract, then -

(a) if the unpaid balance of the Final Sum exceeds any retention which the Employer is for the time being entitled to retain the Employer shall forthwith pay the excess to the Contractor; and

(b) if the total amount paid to the Contractor exceeds the Final Sum the Contractor shall forthwith pay the excess to the Employer.

(5) If after the end of the Maintenance Period, or where there is more than one the end of the last Maintenance Period to expire, the PM has certified that the Contractor has complied with Condition 21 (Defects in Maintenance Periods), and the Final Sum has been calculated and agreed or has been treated as having been agreed under paragraph (3), or in default of agreement has been determined in accordance with the Contract, then -

(a) if the Final Sum exceeds the amount previously paid to the Contractor, the Employer shall forthwith pay the excess to the Contractor; or

(b) if the amount previously paid to the Contractor exceeds the Final Sum, the Contractor shall forthwith pay the excess to the Employer.

50

Certifying payments

(1) Subject to Conditions 59 (Adjudication) and 60 (Arbitration and choice of law), the PM shall certify in the prescribed form to the Employer, with a copy to the Contractor, all net sums (taking into account retention and all set-off or abatement to which the Employer is entitled, but exclusive of VAT) to which the Contractor is entitled.

(2) Where Condition 48(2)(a) (Alternative A (Stage Payment Chart) or Alternative C (Valuation)) applies, the first such certificate shall be issued on a date to be agreed by the Employer and the Contractor, not later than 28 Days after the commencement of the execution of the Works, and subsequent certificates shall be issued on the equivalent date of each subsequent month, if any sum is to be certified.

(3) Where Condition 48(2)(a) (*Alternative B* (Milestone Payment Chart)) applies, certification under paragraph (2) shall apply in respect of the Design Fee and the execution of any other part of the Works Instructed under Condition 10B (Design process and Contract Sum), and each such certificate in respect of the achievement of a Milestone shall be issued within 21 Days of the PM's receipt of an application for payment in respect of the achievement of a Milestone under that Alternative, including (if the Abstract of Particulars so specifies) all amounts due under Condition 48(2)(b)-(h). If the Abstract of Particulars does not so specify, certification under paragraph (2) shall apply in respect of such amounts.

(4) Without prejudice to Condition 1(4)(c) (Definitions, etc.), if any date upon which a certificate should be issued is a Saturday or Sunday, or Christmas Day, Good Friday or a day which under the Banking and Financial Dealings Act 1971 is a bank holiday in England and Wales or, as the case may be, in Scotland or Northern Ireland, the relevant certificate shall be issued on the next working day.

(5) The Contractor shall immediately further copy each such certificate to each of his subcontractors and suppliers.

(6) After certification, the Contractor shall submit to the Employer a VAT invoice for the precise certified sum, plus applicable VAT. Subject to Condition 50A (Payment notification and withholding payment), the Employer shall pay to the Contractor each certified sum, plus applicable VAT, within

30 Days of the Employer's receipt of an invoice, strictly in accordance with this Condition, in respect of that certified sum.

(7) Any certificate may be modified or corrected by any subsequent certificate or by the final certificate for payment. No certificate of the PM shall of itself be conclusive evidence that any work or Things to which it relates are in accordance with the Contract.

50A

Payment notification and withholding payment

(1) The Employer or the Contractor, as the case may be, shall give notice not later than 5 Days after the date on which a payment becomes due from him under the Contract, or would have become due if -

(a) the other party had carried out his obligations under the Contract; and

(b) no set-off or abatement was permitted by reference to any sum claimed to be due under one or more other contracts;

specifying the amount (if any) of the payment made or proposed to be made, and the basis on which that amount was calculated.

(2) Neither the Employer nor the Contractor may withhold payment, after its due date for payment (which shall be its final date for payment for all purposes), of a sum due under the Contract, unless he has given an effective notice of intention to withhold payment. The notice mentioned in paragraph (1) may suffice as a notice of intention to withhold payment if it complies with the requirements of this paragraph. To be effective, a notice of intention to withhold payment must specify -

(a) the amount proposed to be withheld, and the ground for withholding payment; or

(b) if there is more than one ground, each ground and the amount attributable to it;

and must be given not later than 7 Days before the due date for payment of the relevant sum.

51

Recovery of sums

Without prejudice and in addition to any other rights and remedies of the Employer, whenever under or in respect of the Contract, or under or in respect of any other contract between the Contractor or any other member of the Contractor's Group and the Employer or any other member of the Employer's Group, any sum of money shall be recoverable from or payable by the Contractor or any other member of the Contractor's Group by or to the Employer or any other member of the Employer's Group, it may be deducted by the Employer from any sum or sums then due or which at any time thereafter may become due to the Contractor or any other member of the Contractor's Group under or in respect of the Contract, or under or in respect of any other contract between the Contractor or any other member of the Contractor's Group and the Employer or any other member of the Employer's Group. Without prejudice and in addition to any other rights and remedies of the Contractor, each member of the Contractor's Group shall have rights reciprocal to those of each member of the Employer's Group under this Condition.

PARTICULAR POWERS AND REMEDIES

52
Suspension for non-payment

(1) Where a sum due under the Contract (as determined by agreement between the parties, certification and invoicing, adjudication, arbitration or litigation) is not paid in full by the final date for payment and no effective notice to withhold payment has been given under Condition 50A (Payment notification and withholding payment), the person to whom the sum is due has the right (without prejudice to any other right or remedy) to suspend performance of his obligations under the Contract to the party by whom payment ought to have been made ('the party in default').

(2) The right may not be exercised without first giving to the party in default at least 7 Days' notice of intention to suspend performance, stating the ground or grounds on which it is intended to suspend performance.

(3) The right to suspend performance ceases when the party in default makes payment in full of the amount due.

(4) Any period during which performance is suspended in pursuance of the right conferred by this Condition shall be disregarded in computing, for the purposes of any contractual time limit, the time taken, by the party exercising the right or by a third party, to complete any work directly or indirectly affected by the exercise of the right. Where the contractual time limit is set by reference to a date rather than a period, the date shall be adjusted accordingly.

53
Non-compliance with Instructions

If, after receipt of a notice from the PM requiring compliance with any Instruction within a period specified in the notice, the Contractor fails to comply, the Employer may, without prejudice to the exercise of his powers to determine the Contract, provide labour and/or any Things (whether or not for incorporation), or enter into a contract for the execution of any work which may be necessary to give effect to that Instruction. Any reasonable costs and expenses incurred by the Employer over and above those which would have been incurred had the Contractor complied promptly with the Instruction, shall be recoverable by the Employer from the Contractor.

54
Emergency work

(1) If the Contractor is unable or unwilling to carry out promptly any emergency work required by the PM, the Employer may make arrangements for that work to be carried out. If the work carried out by the Employer shall be such as the Contractor is liable under the Contract to carry out or execute at his own expense then the Contractor shall reimburse -

(a) any costs reasonably incurred by the Employer under this paragraph; and

(b) any loss suffered by the Employer because the Contractor has not carried out the work.

(2) In this Condition 'emergency work' means any work which becomes necessary during the execution of the Works or during any Maintenance Period -

(a) to prevent, or alleviate the effects of, any accident, failure or other event in connection with the performance of the Works;

(b) to secure the Works, the Site or any adjoining property from damage; or

(c) without prejudice to Condition 19 (Loss or damage), to repair any damaged or dangerous part of the Works.

55
Liquidated damages

(1) This Condition applies where a rate of liquidated damages for any delay in the completion of the Works or a Section has been specified in the Abstract of Particulars.

(2) If the Works or a Section are or is not completed by the relevant Date for Completion, the Contractor shall immediately become liable to pay to the Employer liquidated damages at the rate specified in the Abstract of Particulars for the period that the Works or any relevant Section remain or remains uncompleted.

(3) Subject to Condition 50A (Payment notification and withholding payment), the Employer may deduct any amount of liquidated damages to which he may be entitled under this Condition from any advances to which the Contractor may otherwise be entitled under Condition 48 (Advances on account).

(4) If the sum due as liquidated damages exceeds any advance payable to the Contractor under Condition 48 (Advances on account), the Contractor shall pay to the Employer the difference. That sum shall be recoverable in accordance with Condition 51 (Recovery of sums).

(5) No payment or concession to the Contractor, or Instruction or VI at any time given to the Contractor (whether before or after the Date or Dates for Completion), or other act or omission by or on behalf of the Employer, shall in any way affect the rights of the Employer to deduct or recover liquidated damages, or shall be deemed to be a waiver of the right of the Employer to recover such damages. The rights of the Employer to deduct or recover liquidated damages may be waived only by notice from the Employer to the Contractor.

56
Determination by Employer

(1) Without prejudice to any other power of determination, the Employer may determine the Contract by notice to the Contractor if -

(a) any ground mentioned in subparagraphs (6) (a), (b) or (e) has arisen; the Employer has given notice to the Contractor specifying the relevant ground and facts; and such ground was in existence 14 Days after such notice was given; or has arisen again at any subsequent time; or

(b) any ground mentioned in paragraph (6) has arisen, other than those mentioned in subparagraphs (6) (a), (b) and (e).

(2) The Employer shall specify in a notice of determination under paragraph (1) which of the grounds mentioned in paragraph (6) apply.

(3) The Employer may, after giving notice of determination under paragraphs (1) or (8), give directions in relation to the performance or completion of any work and any other matters connected with the Works, the Site and any other contract or subcontract.

(4) Any directions under paragraph (3) shall be given not later than three months from the date of the notice of determination under paragraphs (1) or (8) or the relevant Date for Completion, whichever is the sooner.

(5) The Contractor shall comply promptly with any directions given by the Employer under paragraph (3). The Contractor shall be paid for any work so performed as if the directions were Instructions.

(6) The grounds referred to in paragraph (1) are -

(a) the failure of the Contractor to comply with an Instruction within a reasonable period of its issue;

(b) the failure of the Contractor to execute work in a workmanlike or proper manner, or to proceed regularly and diligently with the Works, or the suspension by the Contractor of the Works (otherwise than in accordance with Condition 52 (Suspension for non-payment)), so that in the opinion of the PM the Contractor has not completed or will be unable to secure the completion of the Works or any relevant Section by the Dates for Completion;

(c) where the Contractor is an individual, the insolvency of that individual or, where the Contractor is a partnership, the insolvency of any partner; in this subparagraph 'insolvency' shall include the presentation of a bankruptcy petition where the petitioner is the debtor; the making of a bankruptcy order; the appointment of an interim receiver; the issue of proposals by the debtor to creditors for any arrangement or composition with creditors (whether as a voluntary arrangement under Part VIII of the Insolvency Act 1986 or of the Insolvency (Northern Ireland) Order 1989, or otherwise) or for a conveyance or assignment for the benefit of creditors (whether under the Deeds of Arrangement Act 1914 or otherwise); where the Contractor is a partnership, insolvency shall also include the presentation of a petition by the members of the partnership to wind up the partnership as an unregistered company under Part V of the Insolvency Act 1986 or Part VI of the Insolvency (Northern Ireland) Order 1989; the making of an order to wind up the partnership as an unregistered company; the presentation by the members of the partnership of a petition for the making of an administration order in respect of the partnership; the making of an administration order; or the issue of proposals by the members of the partnership to its creditors for a voluntary arrangement; in Northern Ireland, the term 'insolvency' shall also include in respect of the Contractor, if an individual, or the partnership or any partner if the Contractor is a partnership, becoming insolvent within the meaning of Article 12 of the Construction Contracts (Northern Ireland) Order 1997; and in Scotland, the term 'insolvency' shall also include in respect of the Contractor, if an individual, or the partnership or any partner if the Contractor is a partnership, becoming bankrupt, or having his or its estate sequestrated, or becoming apparently insolvent as defined in the Bankruptcy (Scotland) Act 1985, or entering into a trust deed for his or its creditors, or making a composition or arrangement with his or its creditors;

(d) where the Contractor is a company then (in respect of that company or any company which is for the time being a holding company of the Contractor) the presentation by the company or its directors of a petition for winding-up, or the passing of any resolution for the winding-up, of the relevant company (except for the purposes of amalgamation or reconstruction while solvent); the making by the court of a winding-up order; the appointment of a provisional liquidator; the presentation by the company or its directors of a petition for or the making of an administration order, or a railway administration order; the issue by the company of proposals to creditors for the making of any arrangement or composition with creditors (whether as a voluntary arrangement under Part I of the Insolvency Act 1986, or Part II of the Insolvency (Northern Ireland) Order 1989, or a scheme of arrangement under Section 425 of the Companies Act 1985, or Article 418 of the Companies (Northern Ireland) Order 1986, or otherwise); the appointment of an administrative receiver or a receiver or a receiver and manager in respect of the company or of any of its assets; or the relevant company becoming insolvent within the meaning of Article 12 of the Construction Contracts (Northern Ireland) Order 1997;

(e) failure by the Contractor to comply with Condition 26 (Site admittance) where the Employer determines that such failure is prejudicial to the interests of the Employer or any other member of the Employer's Group;

(f) either of the situations described in Condition 24(3)(a) or (b) (Corruption);

(g) failure by the Contractor to comply with Condition 66 (Performance bond) (if applicable) or Condition 67 (Parent company guarantee) (if applicable), in respect of each of which Conditions time shall be of the essence; or

(h) any breach of the conditions set out in the invitation to tender relating to the Contract.

(7) All Things not for incorporation which are brought onto the Site at the Contractor's expense shall (whether damaged or not) re-vest in and be removed by him as and when they cease to be required in connection with any directions given by the Employer under paragraph (3). From the date of determination the Employer shall be under no liability to the Contractor in respect of any loss or damage to any such Things caused by any of the Accepted Risks.

(8) Without prejudice to any other power of determination, the Employer may at will determine the Contract by notice to the Contractor. If the Employer purports to determine the Contract under paragraph (1), but no ground for determination under paragraph (1) had then arisen, or any such ground which had arisen had been waived by the Employer, the Employer shall be deemed to have decided to determine, and to have determined, the Contract under this paragraph.

57

Consequences of determination by Employer

(1) If the Employer shall determine the Contract for any reason mentioned in Condition 56(6) (Determination by Employer), the following provisions shall apply -

(a) all sums of money that may then be due or accruing due from the Employer to the Contractor shall cease to be due or to accrue due;

(b) the Employer may hire any person, employ other contractors, use any Things on the Site, and may purchase or do anything necessary for the completion of the Works, and the Contractor shall have no claim whatsoever in respect of any such action by the Employer;

(c) the Contractor shall (except where determination occurs by reason of any of the circumstances described in Condition 56(6)(c) and (d) (Determination by Employer)) assign to the Employer, without further payment, the benefit of any subcontract or contract for the supply of any Thing for incorporation which he may have made in connection with the Contract;

(d) the Employer may pay to any subcontractor or supplier any amount due to him which the PM certifies as included in any previous advance to the Contractor, and the amount so paid shall be forthwith recoverable by the Employer from the Contractor; and

(e) the QS shall ascertain and the PM shall certify the cost to the Employer of completion of the Works.

(2) If the total of the following sums (hereafter called 'the first amount') exceeds the total of all the advances paid to the Contractor (or to which he is entitled) under Condition 48 (Advances on account) to the date of determination (hereafter called 'the second amount'), the Employer shall, subject to paragraph (3), hold the amount of the excess. If the second amount exceeds the first amount the Contractor shall be liable to pay the Employer the amount of the excess. The individual sums are -

(a) the value of all the work carried out in accordance with the Contract up to the date of determination;

(b) the value of any work carried out or other things done in accordance with any direction given under Condition 56(3) (Determination by Employer); and

(c) the value (ascertained on the basis of fair and reasonable prices) of all Things for incorporation brought onto the Site, or in the course of preparation or manufacture off the Site, which the Employer elects to keep.

(3) If the total of the cost of completion as certified under subparagraph (1)(e) and the first amount determined in accordance with paragraph (2) exceeds the sum that would have been payable to the Contractor for due completion of the Works, then the Contractor shall pay the Employer the amount of the excess. If the total of the cost of completion as certified under subparagraph (1)(e) and the first amount determined in accordance with paragraph (2) is less than the sum that would have been payable to the Contractor for due completion of the Works, then the Employer shall pay the Contractor the amount of the shortfall.

(4) If the Employer shall determine, or shall be deemed to have determined, the Contract at will under Condition 56(8) (Determination by Employer), Condition 58(5) and (6) (Determination by Contractor) shall apply, as if the Contractor had determined the Contract under that Condition.

58
Determination by Contractor

(1) Without prejudice to any other power of determination, the Contractor may determine the Contract by notice to the Employer if -

(a) the ground mentioned in subparagraph (3)(a) has arisen; the Contractor has consequently suspended performance of his obligations under the Contract under Condition 52 (Suspension for non-payment) for a continuous period of not less than 30 Days; following which period the Contractor has given notice to the Employer specifying the relevant ground and facts; such ground was in existence 14 Days after such notice was given; and the above suspension is still continuing, and has continued without a break since its commencement; or

(b) any ground mentioned in subparagraphs (3)(b) or (e) has arisen; the Contractor has given notice to the Employer specifying the relevant ground and facts; and such ground was in existence 14 Days after such notice was given; or has arisen again at any subsequent time; or

(c) any ground mentioned in paragraph (3) (c) or (d) has arisen;

(2) The Contractor shall specify in a notice of determination which of the grounds mentioned in paragraph (3) apply.

(3) The grounds referred to in paragraph (1) are -

(a) the failure of the Employer to pay any sum certified by the PM for payment to the Contractor (and an appropriate sum in respect of VAT under Condition 45 (VAT)) within the time allowed under Condition 50 (Certifying payments);

(b) the Employer obstructing, or interfering with, the issue of any certificate of the PM referred to in Condition 50(1) (Certifying payments);

(c) where the Employer is an individual, the insolvency of that individual or, where the Employer is a partnership, the insolvency of any partner; in this subparagraph insolvency shall include the occurrence in respect of the individual or partner of any of the matters referred to in Condition 56(6)(c) (Determination by Employer);

(d) where the Employer is a company, the occurrence in respect of that company or any company which is for the time being a holding company of the Employer of any of the matters referred to in Condition 56(6)(d) (Determination by Employer); or

(e) the suspension of the execution of the whole or substantially the whole of the Works (other than the making good of defects in the Works under Condition 21 (Defects in Maintenance Periods) for a continuous period of 182 Days (or such other period as is stated in the Abstract of Particulars)) by reason of -

(i) any Instruction, except where the Instruction was necessary because of any default or neglect by the Contractor or by any employee, agent or subcontractor of his; or

(ii) any of the matters referred to in Condition 46(1)(a), (b) or (c) (Prolongation and disruption).

(4) If the Contractor shall determine the Contract as mentioned in this Condition, or if the Employer shall determine, or shall be deemed to have determined, the Contract at will under Condition 56(8) (Determination by Employer) and Condition 57(4) (Consequences of determination by Employer) consequently applies, the following provisions shall apply.

(5) If the total of the following sums (hereafter called 'the third amount') exceeds the second amount determined in accordance with Condition 57(2) (Consequences of determination by Employer), the Employer shall be liable to pay the Contractor the amount of the excess. If the second amount exceeds the third amount the Contractor shall be liable to pay the Employer the amount of the excess. The individual sums are -

(a) the value of all the work carried out in accordance with the Contract up to the date of determination;

(b) the value of any work carried out or other things done in accordance with any direction given under Condition 56(3) (Determination by Employer);

(c) the value (ascertained on the basis of fair and reasonable prices) of all Things for incorporation brought onto the Site, or in the course of preparation or manufacture off the Site, which the Employer elects to keep;

(d) any reasonable sum expended by the Contractor because of the determination of the Contract in respect of -

(i) the uncompleted part of any subcontract and other contracts (including those for the hire of plant, services and insurance); and

(ii) any unavoidable contract of employment, entered into in connection with the Contract; and

(e) the Contractor's other unavoidable losses or expense directly due to the determination (including, without limitation, loss of profit on the Contract).

(6) All Things not for incorporation which are brought onto the Site at the Contractor's expense shall (whether damaged or not) re-vest in and be removed by him. From the date of determination the Employer shall be under no liability to the Contractor in respect of any loss or damage to any such Things caused by any of the Accepted Risks.

58A

Determination following suspension of Works

(1) Without prejudice to any other power of determination, either the Employer or the Contractor may determine the Contract by notice to the other of them in the event of the suspension of the execution of the whole or substantially the whole of the Works (other than the making good of defects in the Works under Condition 21 (Defects in Maintenance Periods)) for a continuous period

of 182 Days (or such other period as is stated in the Abstract of Particulars): provided that such suspension for that period was by reason of any circumstances, not arising because of any act, neglect or default of the Contractor or of any employee, agent or subcontractor of his; and not arising because of any act, neglect or default of the Employer or the PM or of any other contractor or agent of the Employer; and other than weather conditions; which are outside the control of the Employer and of the Contractor or any of his subcontractors and which could not have been reasonably contemplated under the Contract (including, without limitation, loss or damage to the Works or any Things on the Site caused by an Accepted Risk).

(2) The party giving notice of determination under paragraph (1) shall specify in the notice the grounds for such determination.

(3) If either the Employer or the Contractor shall determine the Contract as mentioned in this Condition, then Condition 58(5) and (6) (Determination by Contractor) shall apply, provided that the Contractor shall not be entitled to any loss of profit on the Contract.

59
Adjudication

(1) The Employer or the Contractor may give notice at any time to the other of his intention to refer a dispute, difference or question arising under, out of, or relating to, the Contract ('a dispute') to adjudication. Within 7 Days of such notice, the Employer or the Contractor may by notice of referral to the adjudicator specified in the Abstract of Particulars refer such dispute to him for adjudication.

(2) The notice of referral shall specify the matter in dispute and shall set out the principal facts and arguments relating to it. Copies of all relevant documents in the possession of the party giving the notice of referral shall be enclosed with the notice of referral. A copy of the notice of referral and enclosures shall be sent by the party giving the notice of referral at the same time to the PM, the QS and the other party.

(3) (a) If the person named in the Abstract of Particulars is unable to act as adjudicator, or is not or ceases to be independent of the Employer, the Contractor, the PM and the QS, he shall be substituted as provided in the Abstract of Particulars.

(b) It shall be a condition precedent to the appointment of an adjudicator that he shall notify both parties that he will comply with this Condition and its time limits.

(c) The adjudicator, unless already appointed, shall be appointed within 7 Days of the giving of a notice of intention to refer a dispute to adjudication under paragraph (1). The Employer and the Contractor shall jointly proceed to use all reasonable endeavours to complete the appointment of the adjudicator and named substitute adjudicator. If either or both such joint appointments has not been completed within 28 Days of the acceptance of the tender, either the Employer or the Contractor alone may proceed to complete such appointments. If it becomes necessary to substitute as adjudicator a person not named as adjudicator or substitute adjudicator in the Abstract of Particulars, the Employer and Contractor shall jointly proceed to use all reasonable endeavours to appoint the substitute adjudicator. If such joint appointment has not been made within 28 Days of the selection of the substitute adjudicator, either the Employer or Contractor alone may proceed to make such appointment. For all such appointments, the form of adjudicator's appointment prescribed by the Contract shall be used, so far as is reasonably practicable. A copy of each such appointment shall be supplied to each party. No such appointment shall be amended or replaced without the consent of both parties.

(4) The PM, the QS and the other party may submit representations to the adjudicator not later than 7 Days from the receipt of the notice of referral.

(5) The adjudicator shall notify his decision to the PM, the QS, the Employer and the Contractor not earlier than 10 and not later than 28 Days from receipt of the notice of referral, or such longer period as is agreed by the Employer and the Contractor after the dispute has been referred. The adjudicator may extend the period of 28 Days by up to 14 Days, with the consent of the party by whom the dispute was referred. The adjudicator's decision shall nevertheless be valid if issued after the time allowed. The adjudicator's decision shall state how the cost of the adjudicator's fee or salary (including overheads) shall be apportioned between the parties, and whether one party is to bear the whole or part of the reasonable legal and other costs and expenses of the other, relating to the adjudication.

(6) The adjudicator may take the initiative in ascertaining the facts and the law, and the Employer and the Contractor shall enable him to do so. In coming to a decision the adjudicator shall have regard to how far the parties have complied with any procedures in the Contract relevant to the matter in dispute and to what extent each of them has acted promptly, reasonably and in good faith. The adjudicator shall act independently and impartially, as an expert adjudicator and not as an arbitrator. The adjudicator shall have all the powers of an arbitrator acting in accordance with Condition 60 (Arbitration and choice of law), and the fullest possible powers to assess and award damages and legal and other costs and expenses; and, in addition to, and notwithstanding the terms of, Condition 47 (Finance charges), to award interest. In particular, without limitation, the adjudicator may award simple or compound interest from such dates, at such rates and with such rests as he considers meet the justice of the case -

(a) on the whole or part of any amount awarded by him, in respect of any period up to the date of the award;

(b) on the whole or part of any amount claimed in the adjudication proceedings and outstanding at the commencement of the adjudication proceedings but paid before the award was made, in respect of any period up to the date of payment;

and may award such interest from the date of the award (or any later date) until payment, on the outstanding amount of any award (including any award of interest and any award of damages and legal and other costs and expenses).

(7) Subject to the proviso to Condition 60(1) (Arbitration and choice of law), the decision of the adjudicator is binding until the dispute is finally determined by legal proceedings, by arbitration (if the Contract provides for arbitration, or the parties otherwise agree to arbitration), or by agreement: and the parties do not agree to accept the decision of the adjudicator as finally determining the dispute.

(8) In addition to his other powers, the adjudicator shall have power to vary or overrule any decision previously made under the Contract by the Employer, the PM or the QS, other than decisions in respect of the following matters -

(a) decisions by or on behalf of the Employer under Condition 26 (Site admittance);

(b) decisions by or on behalf of the Employer under Condition 27 (Passes) (if applicable);

(c) provided that the circumstances mentioned in Condition 56(1)(a) or (b) (Determination by Employer) have arisen, and have not been waived by the Employer, decisions of the Employer to give notice under Condition 56(1)(a), or to give notice of determination under Condition 56(1);

(d) decisions or deemed decisions of the Employer to determine the Contract under Condition 56(8) (Determination by Employer);

(e) provided that the circumstances mentioned in Condition 58A(1) (Determination following suspension of Works) have arisen, and have not been waived by the Employer, decisions of the Employer to give notice of determination under Condition 58A(1); and

(f) decisions of the Employer under Condition 61 (Assignment).

In relation to decisions in respect of those matters, the Contractors's only remedy against the Employer shall be financial compensation.

(9) Notwithstanding Condition 60 (Arbitration and choice of law), the Employer and the Contractor shall comply forthwith with any decision of the adjudicator; and shall submit to summary judgment and enforcement in respect of all such decisions.

(10) If requested by one of the parties to the dispute, the adjudicator shall provide reasons for his decision. Such requests may only within 14 Days of the decision being notified to the requesting party

(11) The adjudicator is not liable for anything done or omitted in the discharge or purported discharge of his functions as adjudicator, unless the act or omission is in bad faith. Any employee or agent of the adjudicator is similarly protected from liability.

60

Arbitration and choice of law

(1) The procedures for arbitration set out in this Condition shall be utilised by the Employer and the Contractor with regard to disputes, differences or questions between the Employer and the Contractor arising out of, or relating to, the Contract, whether during the course or after the determination thereof (other than disputes, differences or questions arising out of, or relating to, the enforceability or enforcement of any adjudicator's decision; or Condition 45 (VAT); or the statutory tax deduction scheme under Part XIII (Miscellaneous Special Provisions) Chapter IV (Sub-Contractors in the Construction Industry) of the Income and Corporation Taxes Act 1988). The dispute, difference or question shall after notice by either party to the other be referred to the single arbitrator specified in the Abstract of Particulars. If the person named in the Abstract of Particulars is deceased or unwilling or unable to act as arbitrator, or is not or ceases to be independent of the Employer, the Contractor, the PM and the QS, he shall be substituted as provided in the Abstract of Particulars. It shall be a condition precedent to the appointment of an arbitrator that he shall notify both parties that he will forthwith commence his duties, and will comply with this Condition and its time limits. In addition to his other powers, the arbitrator shall have the fullest possible powers -

(a) to rectify the Contract;

(b) to order inspections, measurements and valuations;

(c) to vary or overrule any decision previously made under the Contract by the Employer, the PM, the QS or an adjudicator, provided that the Contractor's only remedy against the Employer in relation to decisions in respect of the matters listed in Condition 59(8)(a)-(f) (Adjudication) shall be financial compensation;

(d) to order consolidation of the proceedings with other proceedings; and/or that concurrent hearings shall be held; and to make such orders and directions relating to such consolidation and hearings as he thinks fit; and

(e) to make orders, directions and awards in the same way as if all the procedures of the High Court of Justice in England (if the proper law of the Contract is English law), or the High Court of Justice in Northern Ireland (if the proper law of the Contract is Northern Ireland law), or the Court of Session in Scotland (if the proper law of the Contract is Scots law), including, without limitation, as to joining one or more defendants or defenders or joining co-defendants or co-defenders or third parties, were available to the parties and to the arbitrator.

Provided always, that where any dispute, difference or question has been referred to an adjudicator under Condition 59 (Adjudication), and the adjudicator has issued his decision thereon, a party shall not be entitled to refer such dispute, difference or question to arbitration, and the adjudicator's decision thereon shall become unchallengeable, unless that party serves the above notice within 56 Days of receipt of notification of the adjudicator's decision: and, for the avoidance of doubt, this proviso shall apply whether or not the adjudicator has notified his decision within the time limit specified in Condition 59(5).

(2) Unless the parties otherwise agree -

(a) no reference shall be made under paragraph (1) until after the completion, alleged completion or abandonment of the Works, or the determination of the Contract;

(b) no reference shall be made under paragraph (1) in respect of a dispute, difference or question which has been referred to an adjudicator under Condition 59 (Adjudication), until the notification by the adjudicator of his decision thereon; or the expiry of 28 Days from the receipt by the adjudicator of the notice of referral under Condition 59(1), or such longer time as is allowed for the adjudicator's decision in accordance with Condition 59; whichever is the earlier;

(c) the arbitrator shall hold a preliminary meeting with the parties forthwith on his acceptance of office and will fix a timetable for the delivery of points of claim, defence or counter-claim and any other pleadings, for the discovery and inspection of documents, for the inspection of the Works (if necessary) and for the hearing of any oral evidence (if necessary), which timetable shall not without the consent of the parties exceed a period of 6 months from the date of that preliminary meeting;

(d) the parties shall ensure that any evidence whether oral or written and any document or argument required to be submitted to the arbitrator is submitted to him in accordance with the timetable; and

(e) the arbitrator shall give his award not later than 3 months from the end of the period mentioned in subparagraph (c).

(3) (a) If the Works are in England or Wales, the proper law of the Contract shall be English law, a reference to arbitration in accordance with this Condition shall be a reference to which the Arbitration Act 1996 applies, and that Act shall have effect subject to the provisions of this Condition.

(b) If the Works are in Scotland, the Contract shall in all respects be construed and operated as a Scottish contract, and shall be interpreted in accordance with Scots law.

(c) If the Works are in Northern Ireland, the proper law of the Contract shall be Northern Ireland law, a reference to arbitration in accordance with this Condition shall be a reference to which the Arbitration Act 1996 applies, and that Act shall have effect subject to the provisions of this Condition. Whatever the nationality, residence or domicile of the Contractor, any sub-contractor or supplier, or the arbitrator, such arbitration shall be conducted solely within Northern Ireland.

(4) If the Contract is subject to Scots law, the following shall apply -

(a) References throughout the Contract to a 'Deed' or 'Deeds' shall be construed as references to a document or documents subscribed in accordance with the requirements for a self-proving document under the Requirements of Writing (Scotland) Act 1995.

(b) Condition 48C (Payment for Things off-Site) shall not apply.

(c) In Condition 59(9) (Adjudication), the words 'summary judgement and' shall be deleted: and there shall also be added at the end the following sentence:

'Where the Employer, the Contractor or the adjudicator wishes to register the decision of the adjudicator for execution in the Books of Council and Session, any other party shall, on being requested to do so, forthwith consent to such registration by subscribing the decision before a witness'.

(d) In subparagraph (2)(c), the words 'points of claim' shall be substituted by the words 'the statement of claim'; and the word 'discovery' shall be substituted by the word 'recovery'.

(e) The law of Scotland shall apply to any adjudication and to any arbitration, and any arbitration shall be conducted in accordance with the Scottish Arbitration Rules of the Chartered Institute of Arbitrators (Scottish Branch) 1998 Edition, or such amended Scottish Rules as the said Chartered Institute may have adopted to take effect before the commencement of the arbitration, the terms of which are deemed to be incorporated herein, provided that to the extent that there is inconsistency between the terms of those Rules and this Condition, the latter shall have precedence.

(f) Subject to the provisions of the Contract in respect of adjudication and arbitration, the parties submit to the non-exclusive jurisdiction of the Scottish courts.

ASSIGNMENT, SUBLETTING, SUBCONTRACTING, SUPPLIERS AND OTHER WORKS

61

Assignment

The Contractor shall not, without the consent in writing of the Employer, assign or transfer the Contract, or any part, share or interest under it. No sum of money to become payable under the Contract shall be payable to any person other than the Contractor without the Employer's written consent. The Employer may assign or transfer the benefit of the Contract, or any part, share or interest under it. If so provided by the Abstract of Particulars, this power shall only be exercisable after certification by the PM under Condition 39 (Certifying completion) of the completion of the Works or the last Section thereof in respect of which completion is certified, or the determination of the Contract for any reason whatsoever, including (without limitation) breach by the Employer, whichever is the earlier. Neither party may assign or transfer the whole or any part of the burden of the Contract.

62

Subletting

(1) (a) Except where the Employer has accepted a subletting proposal prior to the award of the Contract, or the Contract specifies the subletting of work, the Contractor shall not sublet any part of the Contract without the prior consent of the PM.

(b) The Contractor shall provide such details of any subcontractor he wishes to engage, and (other than rates and prices) of any subcontract entered into, as the PM may require.

(2) The Contractor shall ensure that each subcontract entered into will enable him to fulfil his obligations under the Contract. No subcontract shall include any provision making payment under the subcontract conditional on the payer receiving payment from a third person unless that third person, or any other person payment by whom is under the subcontract (directly or indirectly) a condition of payment by that third person, is insolvent within the meaning of Section 113 of the Housing Grants, Construction and Regeneration Act 1996, or, if the Works are in Northern Ireland, of Article 12 of the Construction Contracts (Northern Ireland) Order 1997. Each subcontract shall include -

(a) Power to determine the subcontract as a consequence of determination of this Contract under Conditions 34A (Deemed determination), 56 (Determination by Employer), 57 (Consequences of determination by Employer), 58 (Determination by Contractor) or 58A (Determination following suspension of Works).

(b) A provision to the effect that from the commencement to the completion of the subcontract work all Things belonging to the person who enters into the subcontract which are brought on the Site in connection with the subcontract shall vest in the Contractor subject to any right of the Contractor to reject the same.

(c) Such provisions as may be necessary to enable the Contractor to fulfil his obligations to the Employer under the Contract.

(d) Such provisions as will impose on the person who enters into the subcontract liabilities similar to those imposed on the Contractor under the Contract.

(e) A provision to the effect that no part of the subcontract work shall be further sublet without the consent of the Contractor.

(f) (Not used)

(g) Provisions equivalent to Conditions 1A (Fair dealing and teamworking), 50A (Payment notification and withholding payment), 52 (Suspension for non-payment) and 59 (Adjudication).

(h) Provisions entitling the subcontractor to payment by instalments, stage payments or other periodic payments for any work under the subcontract unless it is specified in the subcontract that the duration of the work is to be less than 45 Days; or it is agreed between the Contractor and the subcontractor that the duration of the work is estimated to be less than 45 Days; and stipulating the amounts of the payments and the intervals at which, or circumstances in which, they become due.

(i) Terms and conditions providing an adequate mechanism for determining what payments become due under the subcontract, and when; providing for a final date for payment in relation to any sum which becomes due; and stipulating how long the period (not longer than 30 Days) is to be between the date on which a sum becomes due and the final date for payment.

(j) Any subcontract terms and conditions, or form of subcontract, prescribed in the Abstract of Particulars.

(3) Without prejudice to the obligations of the Contractor under any of the provisions of the Contract, the Contractor shall, whenever requested to do so by the Employer, take any necessary action to ensure that a person who has entered into a subcontract complies with and performs all obligations imposed upon him.

(4) Where for any reason a subcontract is determined or assigned because of the default or failure of the subcontractor, the Contractor shall at his own expense secure completion of the subcontract work.

(5) The Contractor shall be responsible for any subcontractor or supplier employed by him in connection with the Works, whether or not approved by the Employer or the PM, or appointed by the Contractor in accordance with an Instruction or otherwise.

(6) The Contractor shall make good any loss suffered or expense reasonably incurred by the Employer by reason of any default or failure, whether total or partial, on the part of any subcontractor or supplier.

63

(Not used)

63A

(Not used)

64

Provisional sums

The full amount of any provisional lump sums included in the Contract shall be deducted from the Contract Sum, and the value of work ordered and executed thereunder shall be ascertained as provided by Condition 42 (Valuation of Variation Instructions). No work under these items is to be commenced without Instruction from the PM.

65

Other works

(1) The Employer shall have power at any time to execute other works (whether or not in connection with the Works) on the Site at the same time as the Works are being executed. The Contractor shall give reasonable facilities for these works.

(2) The Contractor shall not be responsible for damage done to other works except for damage caused by the negligence, omission or default of his workpeople, agents or subcontractors. Any damage done to the Works in the execution of other works shall, for the purposes of Condition 19 (Loss or damage), be deemed to be damage which is wholly caused by the neglect or default of the Employer or of any other contractor or agent of the Employer.

PERFORMANCE BOND, PARENT COMPANY GUARANTEE AND COLLATERAL WARRANTIES

66

Performance bond (only applicable if stated in Abstract of Particulars)

The Contractor shall, within 21 Days after the Employer notifies the Contractor when he may take possession of the Site, or those parts of the Site defined in the Contract, pursuant to Condition 34 (Commencement and completion), or within 21 Days after the determination of the Contract Sum pursuant to Condition 10B (Design process and Contract Sum), whichever is later, deliver to the Employer a performance bond in the form prescribed by the Contract from the surety or sureties named in the tender, in an amount of 10% of the Contract Sum (or such other percentage as is stated in the Abstract of Particulars).

67

Parent company guarantee (only applicable if stated in Abstract of Particulars)

The Contractor shall, within 21 Days of the acceptance of the tender, deliver to the Employer a parent company guarantee in the form prescribed by the Contract from its ultimate holding company (if any) named in the tender.

68

Collateral warranties (only applicable if stated in Abstract of Particulars)

(1) (a)The Contractor shall use reasonable endeavours (unless waived by the Employer) to procure that each subcontract or supply contract with each subcontractor or supplier shall contain obligations on the relevant subcontractor or supplier to execute, in favour of the Employer, a Deed in the form prescribed by the Contract, or a similar form reasonably required by the Employer, and deliver the same to the Employer; together in each case (unless waived by the Employer) with a guarantee in the form prescribed by the Contract, or a similar form reasonably required by the Employer, from the ultimate holding company (if any) of the relevant subcontractor or supplier in respect of the subcontractor's or supplier's obligations pursuant to such Deed. The Contractor shall enforce such obligations, or such modified obligations as are referred to below.

(b) If, despite the Contractor having used such reasonable endeavours, the subcontractor or supplier will not accept such obligations, or will only accept them in a modified form, the Contractor shall notify the Employer, who may agree in writing that the relevant subcontract or supply contract need not contain such obligations, or that the relevant obligations may be in a modified form agreeable to the subcontractor or supplier.

(c) Failing such agreement by the Employer, the Contractor shall not enter into a relevant subcontract or supply contract with the relevant subcontractor or supplier.

(2) The above obligations for the provision of Deeds and guarantees shall continue notwithstanding determination of the Contract for any reason whatsoever, including (without limitation) breach by the Employer.

SCHEDULE OF TIME LIMITS

Condition	Subject	Limit	Imposed on
8(4)	Send other party an insurance certificate	**21 days** from tender acceptance or renewal/ expiry date	Insuring Party
10B(2)	Submit to the QS lump sum quotation of proposed Contract Sum	**56 days** before Design Process Event	Contractor
10B(8)	Notify QS of any disagreement with QS's determination of Contract Sum	**14 days** from QS's notification	Contractor
21(4)	Defects which have been made good	Relevant Maintenance Period	Contractor
35(2)	Hold progress meetings	Monthly	PM
35(3)	Submit report to PM prior to each progress meeting	**5 days**	Contractor
35(4)	Issue a statement of progress after each progress meeting	**7 days**	PM
36(1)	Notify Contractor of decisions regarding extension of time for whole or part of Works	**42 days** from date of notice by Contractor	PM
36(4)	Make final decision on all outstanding and interim extensions of time	**42 days** after completion of Works	PM
36(5)	Notify PM of dissatisfaction with decision	**14 days** from receipt of decision	Contractor
36(5)	Notify Contractor of any amendment to decision	**28 days** from receipt of notification	PM
40(3)	Confirm oral Instructions in writing	**7 days**	PM
40(5)	Submit to the QS a lump sum quotation showing the cost of complying with a VI	**21 days** from receipt of VI	Contractor
42(3)	Decide on lump sum quotation	**21 days** from receipt	PM
42(7)	Submit information required by Condition 41(4)	**14 days** from request by QS	Contractor

Condition	Subject	Limit	Imposed on
42(8)	Notify Contractor of valuation of a VI	**28 days** from receipt of information under Condition 42(7)	QS
42(9)	Notify QS of any disagreement with valuation	**14 days** from QS's notification	Contractor
42(12)	Deliver daywork vouchers to QS for valuation	By end of week following that in which work is done	Contractor
43(2)	Submit information to QS for valuation of other Instructions	**28 days** of complying with Instructions	Contractor
43(2)	Notify Contractor of amount determined	**28 days** of receipt of information	QS
43(3)	Notify QS of disagreement with amount determined	**14 days** of receipt	Contractor
46(3)	Provide details of expenses incurred as a result of prolongation and disruption	**56 days** of incurring expense	Contractor
46(5)	Notify Contractor of decision on expenses claimed	**28 days** from receipt of information referred to in Condition 46(3)(b)	QS
48A	Deliver retention payment bond	**28 days** of notice of possession of the Site	Contractor
48B(1)	Deliver mobilisation payment bond	**28 days** of notice of possession of the Site	Contractor
48B(1)	Pay mobilisation payment	**14 days** of delivery of mobilisation payment bond	Employer
49(2)	Send copy of draft final account to Contractor	**6 months** from certified completion	QS
49(2)	Notify agreement or disagreement with draft final account	**3 months** from receipt of draft final account	Contractor
50(2)	Issue certificates for payment	**28 days** from commencement of Works, and then monthly	PM
50(3)	Issue certificates for payment	**21 days** from Contractor's claim	PM
50(6)	Pay certified sum	**30 days** from certification and invoicing	Employer

Condition	Subject	Limit	Imposed on
50A(1)	Notify amount and basis of each payment	**5 days** after due date for payment	Paying party
50A(2)	Notice of intention to withhold payment	**7 days** before due date for payment	Party serving notice
52	Notice of intention to suspend performance	**7 days** before suspension	Party serving notice
56(1)	Warning notice of default	**14 days** from notice	Contractor
56(4)	After determination of the Contract, directions may be given in relation to performance or completion	Not later than **3 months** from notice of determination or Date for Completion, whichever is sooner	Employer
58(1)	Warning notice of default	**14 days** from notice	Employer
59(3)	Appointment of adjudicator and named substitute adjudicator	**28 days** from acceptance of tender	Both parties
59(4)	Submission of representations to the adjudicator	Not later than **7 days** from receipt of notice of referral	Recipient
59(5)	Notification of decision to PM, QS, Employer and Contractor	Not earlier than **10** and not later than **28 days** from receipt of notice of referral	Adjudicator
60(1)	Notice of arbitration	**56 days** after notification of adjudicator's decision	Party serving notice
60(2)(c)	Arbitrator's timetable for delivery of points of claim, etc.	Not (without consent) to exceed **6 months** from date of preliminary meeting	Arbitrator
60(2)(e)	Award to be made by arbitrator	Not later than **3 months** from end of period in Condition 60(2)(c)	Arbitrator
66	Deliver performance bond	**21 days** from notice of possession of the Site	Contractor
67	Deliver parent company guarantee	**21 days** from acceptance of tender	Contractor

NOTE: Printed time limits can be extended by agreement within the terms of Condition 1(4). Days means calendar days - see Condition 1(1).

ALPHABETICAL INDEX

MODEL FORMS

The model forms required for use with *GC/Works/1 Two Stage Design and Build (1999)* are contained on the disk which accompanies this volume and includes the following:

DOCUMENTS COLLATERAL TO THE CONTRACT

01 Abstract of Particulars and Addendum

02 Invitation to Tender and Schedule of Drawings

03 Tender and Tender Price Form

04 Contract Agreement (England, Wales & Northern Ireland)

05 Contract Agreement (Scotland)

06 Insurance Documents (Condition 8)

07 Performance Bond (Condition 66)

08 Parent Company Contract Performance Guarantee (Condition 67)

09 Retention Payment Bond (Condition 48A)

10 Mobilisation Payment Bond (Condition 48B)

11 Subcontractor's Collateral Warranty (Condition 68)

12 Parent Company Guarantee of Subcontractor's Collateral Warranty (Condition 68)

13 Adjudicator's Appointment (Condition 59)

ADMINISTRATIVE DOCUMENTS

14 Notice of Delegation (Condition 4)

15 Notice of Possession (Condition 34)

16 Certificate of Completion (Condition 39)

17 Maintenance Certificate (Conditions 21 and 39)

18 Project Manager's Instruction (Condition 40)

19 Interim Payment Certificate (Conditions 48 and 50)

20 Final Payment Certificate (Conditions 49 and 50)

21 Notice of Intention to Withhold Payment (Condition 50A)

22 Notice of Non-Compliance with Instruction (Condition 53)

23 Employer's Warning Notice (Condition 56)

24 Employer's Notice of Determination (Condition 56)

25 Employer's Notice of Intention to Refer to Adjudication (Condition 59)

26 Employer's Notice of Referral to Adjudication (Condition 59)

STAGE PAYMENT CHART, ETC

27 Example Stage Payment Chart (Condition 48(2), Alternative A)

28 Chart Banding Calculation Sheet (Condition 48(2), Alternative A)

COMMENTARY

This commentary has been prepared as a general guide. **This commentary does not form part of, and shall not affect the interpretation of, any contract.** Some practice and policy context has been included, but this is not the primary aim of the commentary. For further guidance on best practice and policy see, for example:

- Other PACE/CAU publications such as the '*Guide to the Appointment of Contractors and Consultants*', and periodic Information Notes on current issues.
- Central Unit on Procurement guides, and other advice from the Treasury (the Central Unit on Procurement has recently become the Procurement Practice & Development Group of the Treasury).
- Publications of the Construction Industry Board.

Legal advice should be taken if it is proposed to amend any of the General Conditions.

It is important that the appointments of the Employer's construction consultants should appropriately correspond with the Contract, so that they are required to carry out their duties as provided by the Contract.

GC/Works/1 Single Stage Design & Build (1998) is a lump form of contract, which is intended to support a single stage tendering procedure, without a separate design stage. The form of contract is sufficiently flexible to allow for varying amounts of design input from the Contractor. The Employer provides Employer's Requirements, to which the Contractor responds with Contractor's Proposals, and the Contractor develops the detailed design on the basis outlined in those documents.

GC/Works/1 Two Stage Design & Build (1999) is intended to support a two stage procedure, with a separate design stage. It is also a lump sum form of contract, but the lump sum is arrived at in two stages, as explained below.

Naturally, the two documents are very similar. The new features of *GC/Works/1 Two Stage Design & Build (1999)* are as follows:

- At the time the Contract is entered into, the Contract Sum is not ascertained, owing to the lack at that time of design and other information upon which to base the Contract Sum. However, a separate Design Fee, carved out of the Contract Sum, is ascertained at that time, and the Contractor will have submitted a Pricing Document as an aid to the subsequent ascertainment of the Contract Sum.
- A Design Process Event is identified. This consists of two distinct elements:
    - Determination of the Contract Sum during the design stage either by agreement of the parties or, in the absence of agreement, by the QS (subject to later adjudication and arbitration at the instance of the Contractor).
    - Achievement of a certain design progress milestone. This may be completion of the Design, or some earlier design stage, which must be identified in the entry in the Abstract of Particulars relating to the new Condition 10B (Design process and Contract Sum).
- The Contractor proceeds with site and/or soil investigations and the Design, but generally with no other work until given notice of possession of the Site.
- The Contractor is paid the Design Fee monthly, without retention.

- Under Condition 10B(10) (Design process and Contract Sum), the PM certifies when the Design Process Event has been achieved.

- The Employer elects in his absolute discretion whether or not to proceed with the Contract and give the Contractor notice of possession of the Site. If he does not do so within a limited time, the Contract is deemed determined, and the Contractor is entitled to be paid for work done to that time.

- If the Employer elects to let the Contract determine, but later decides to proceed with the Works, he may use the initial Contractor's design, whether or not he further employs that Contractor in connection with the Works.

- If the Employer elects to give notice of possession of the Site, the time for the Contractor to provide the performance bond, retention payment bond and mobilisation payment bond (if any) runs from the giving of that notice.

Significant differences between *GC/Works/1 Single Stage Design & Build (1998)* and *GC/Works/1 Two Stage Design & Build (1999)* are explained in more detail below.

For an explanation of other parts of *GC/Works/1 Two Stage Design & Build (1999)*, see the published commentary on *GC/Works/1 Single Stage Design & Build (1998)*.

Condition 1 (Definitions, etc.)

The definition of the Contract Sum has been amended to refer to 'the sum accepted by the Employer when notifying the Contractor when he may take possession of the Site, or those parts of the Site defined in the Contract, pursuant to Condition 34 (Commencement and completion)', rather than 'the sum accepted by the Employer when awarding the Contract'. This is because usually only the Design Fee, part of the Contract Sum, will be ascertained at the time of entry into the Contract, with the Contract Sum to be determined, during the design stage, under Condition 10B (Design process and Contract Sum). If the Contract Sum is not determined by mutual agreement under that Condition, but unilaterally by the QS as provided by the Condition, the Employer is deemed to accept the QS's determination of the Contract Sum by issuing notice of possession under Condition 34 (Commencement and completion) - see Condition 34(1).

Two definitions have been added - 'the Design Fee' and 'Design Process Event'.

The Design Fee is that part of the Contract Sum, which must be identified as the Design Fee in the Pricing Document, payable to the Contractor for carrying out the design process required under Condition 10B (Design process and Contract Sum). If the Contract is eventually deemed determined under Condition 34A (Deemed determination), the Contractor's entitlement will be to that part of the Design Fee which he has at that point earned, plus payment for any other work Instructed under Condition 10B (Design process and Contract Sum).

The Design Process Event comprises both the determination of the Contract Sum in accordance with Condition 10B (Design process and Contract Sum), and the happening of an event specified for the purpose in the Abstract of Particulars, and shown in the Programme. If the Abstract of Particulars does not define the relevant event, it will be the completion of the Design and the signing and dating of all Design Documents by the PM under Condition 10A(2) (Design Documents). Frequently, the event may be defined in the Abstract of Particulars as some point short of that. This will be some earlier point in the design process at which the Employer considers it safe to proceed with the Works on Site, and so to notify the Contractor when he may take possession of the Site, or those parts of the Site defined in the Contract, pursuant to Condition 34 (Commencement and completion). This point must obviously occur reasonably late in the design process, so that enough detail will be in existence in order realistically to determine the Contract Sum. An example of such a point could be the issue of building regulation approval in respect of the Works. The disadvantage to the Employer of selecting a point in the design process, and determining the Contract Sum, before the completion of the Design, is that later Employer-initiated changes in the Works will usually constitute Variation Instructions, with financial consequences to the Employer.

Condition 10B (Design process and Contract Sum)

This Condition shows how the Design Process Event is to be achieved.

Paragraph (1) shows what the Contractor is entitled and authorised to do upon entry into the Contract. He is to carry out any site and/or soil investigations required by the Contract (which will usually be specified in the Employer's Requirements) and proceed with the Design so that the Design Process Event will be achieved in accordance with the Programme. Unless and until the Employer notifies the Contractor, pursuant to Condition 34 (Commencement and completion), when he may take possession of the Site, or those parts of the Site defined in the Contract, the Contractor is not to execute any part of the Works (except the above site and/or soil investigations and Design), unless so Instructed. Any such additional work Instructed should normally be only of a minor and preparatory nature, because of the possibility that the Contract may later be deemed determined under Condition 34A (Deemed determination).

Paragraphs (2) - (9) deal with the determination of the Contract Sum. These paragraphs are based upon the procedures for the valuation of Variation Instructions, as set out in Conditions 41 (Valuation of Instructions - Principles) and 42 (Valuation of Variation Instructions), with appropriate changes.

Under paragraph (2), as soon as reasonably practicable, and not later than 56 Days before the time shown in the Programme for the certification of the Design Process Event, the Contractor is to submit to the QS a written fully itemised lump sum quotation of the proposed Contract Sum, based as far as possible on the rates and prices contained within the Pricing Document. The Contractor is to include with his quotation such information as will enable the QS to evaluate the quotation, and is also promptly to submit to the QS any information he reasonably requires in explanation and support of the quotation.

Under paragraph (3), the Contract Sum is to be determined either by acceptance by the PM of a lump sum quotation of the proposed Contract Sum prepared by the Contractor and submitted to the QS in accordance with paragraph (2), or such other lump sum as may be agreed by negotiation on such a quotation; or by the QS in accordance with paragraph (6). The ability of the QS unilaterally to determine the Contract Sum in the absence of mutual agreement is essential, in order to avoid the creation of an unenforceable 'agreement to agree'. If the determination of the Contract Sum is left to an unenforceable 'agreement to agree' at the end of the design stage, one might as well not create a two stage design and build Contract at all, but enter initially into a separate design contract with a *GC/Works/1 Single Stage Design & Build (1998)* Contract to be negotiated later.

Under paragraph (4), the PM is to notify the Contractor, as soon as reasonably practicable after the receipt of the lump sum quotation, whether or not it is accepted, or if it is not acceptable, whether he is prepared to agree any other lump sum. If accepted, the aggregate amount specified in that quotation, or otherwise agreed between the Employer and the Contractor, is to be the Contract Sum.

Under paragraph (5), if the Contractor fails to provide a lump sum quotation in accordance with paragraph (2), or the Employer and the Contractor fail to agree the Contract Sum by negotiation, the PM is to instruct the QS to determine the Contract Sum.

Under paragraph (6), where the QS is required to determine the Contract Sum under paragraph (5), he is to do so:

- by measurement and valuation at the rates and prices in the Pricing Document; or
- if that is not possible, then by measurement and valuation as rates and prices deduced or extrapolated from the rates and prices in the Pricing Document; or
- if that is not possible, then by measurement and valuation at fair rates and prices, having regard to current market prices.

Under paragraph (7), the QS, as soon as reasonably practicable, is to notify the Contractor of his determination of the Contract Sum under paragraph (6). Such determination is, subject to paragraph (8), subject to later adjudication under Condition 59 (Adjudication) and arbitration under Condition 60 (Arbitration and choice of law), but only at the instance of the Contractor (see Condition 34(1) (Commencement and completion)).

Under paragraph (8), if the Contractor disagrees with the whole or part of the QS's determination of the Contract Sum, he must, within 14 Days of the QS's notification under paragraph (7), give his reason for disagreement and his own valuation. In any other case, and in all other respects, he is to be treated as having accepted the notification under paragraph (7), and the QS's determination of the Contract Sum.

Under paragraph (9), any percentage or lump sum adjustments made in the pricing of the Pricing Document are to be deemed to be applicable to the determination of the Contract Sum.

Under paragraph (10), the PM is to certify the date when the Design Process Event has been achieved - that is, when both elements of the Design Process Event (the determination of the Contract Sum and the reaching of the specified point in the design process) have been achieved.

Condition 33 (Programme)

Paragraph (1) now refers specifically to the Design Process Event, in accordance with the definition of that term in Condition 1(1) (Definitions, etc.). As the definition of the Works includes the Design, the Programme must also include appropriate details relating to the design process.

Condition 34 (Commencement and completion)

Paragraph (1) now provides that, within the period specified in the Abstract of Particulars, the Employer has an absolute discretion whether or not to proceed with the Contract. Within that period, the Employer must either notify the Contractor that the Employer has decided not to proceed with the Contract, or notify the Contractor when he may take possession of the Site, or those parts of the Site defined in the Contract. Any such notification to the Contractor when he may take possession of the Site, or those parts of the Site defined in the Contract, is to constitute acceptance by the Employer of any then existing or future determination of the Contract Sum by the QS in accordance with Condition 10B (Design process and Contract Sum). This interlocks with the definition of the Contract Sum. The words 'existing or future' are intended to cover the (probably rare) case, referred to below, where the Employer gives notice of possession before achievement of the Design Process Event (the definition of which includes the determination of the Contract Sum). The QS is, after all, one of the Employer's own consultants. In all cases where the Employer has not gone so far as to give notice of possession before determination of the Contract Sum, if he does not like the QS's determination of the Contract Sum, he may determine the Contract under Condition 34A (Deemed determination).

Any date of possession so notified is to be not more than 14 Days after the end of the above period specified in the Abstract of Particulars. The Date or Dates for Completion are to be calculated from the date of possession so notified by the Employer to the Contractor.

The period to be specified in the Abstract of Particulars, within which notice of the date of possession may be given, commences a certain number of Days after acceptance of the tender - this is to allow a reasonable time for the achievement of the Design Process Event. The period ends a certain number of Days after certification by the PM under Condition 10B (Design process and Contract Sum) of achievement of the Design Process Event. The date of possession so notified is not to be more than 14 Days after the end of that period.

Note that the Employer may give notice of possession before achievement of the Design Process Event, provided that the specified number of Days since acceptance of the tender has elapsed. The Employer may do so, for example, in an urgent case where the Design Process Event is obviously imminent.

Condition 34A (Deemed determination)

In this new Condition, paragraph (1) provides that, if the Employer, within the period specified in the Abstract of Particulars, and pursuant to Condition 34 (Commencement and completion), either notifies the Contractor that the Employer has decided not to proceed with the Contract, or does not give notice of possession of the Site, the Contract is automatically determined. Deemed determination will usually occur because the Employer is not keen on the Contractor's design, or decides to postpone or cancel the Works.

Paragraph (2) deals with the consequences of such deemed determination. The Contractor is to cease work and deliver to the Employer all documents and particulars comprising the Design or parts of the Design. He is entitled to be paid any balance of the Design Fee, or of that part of it which relates to work he has executed. He will normally have earned the whole of the Design Fee, as the Design Process Event will have been certified. He is also entitled to be paid for any other work Instructed under Condition 10B (Design process and Contract Sum). Such payments are the Contractor's sole and exclusive entitlement in respect of the deemed determination.

If the Employer later wishes to proceed with the Works, using the initial Contractor's design, he is free to do so, exercising his rights under Conditions 10(4) (Design) and 10A(3) (Design Documents), whether or not he further employs that Contractor in connection with the Works.

Condition 48 (Advances on account)

This Condition has been amended to deal with payment of the Design Fee, and for any other work Instructed under Condition 10B (Design process and Contract Sum).

A new subparagraph (i) has been added to Alternative A (Stage Payment Chart) accordingly. This provides for payment (without retention) of the relevant sums, and for the Contractor to make applications for such payment, supported by valuations. These applications will be monthly - see Condition 50(2) and (3) (Certifying payments). Naturally, no application is necessary under subparagraph (ii) for payment in accordance with the Stage Payment Chart.

A new subparagraph (i) to the same effect has been added to Alternative B (Milestone Payment Chart).

An amendment to the same effect has been made at the beginning of Alternative C (Valuation). That Alternative already provides for Contractor's monthly applications for payment.

Condition 48A (Retention payment bond)

Subparagraph (b) now provides for the retention payment bond to be delivered by the Contractor to the Employer within 28 Days of notice of possession being given under Condition 34 (Commencement and completion), rather than within 28 Days of acceptance of tender. This is because, if notice of possession is not given, the Contract will be deemed determined under Condition 34A (Deemed determination). Normally, it will be only after notice of possession has been given that payments subject to retention will arise.

Condition 48B (Mobilisation payment)

Subparagraph (1)(b) of this optional Condition has been amended to the same effect as Condition 48A(b) (Retention payment bond). The Employer will not normally wish to make a mobilisation payment while there is a possibility that the Contract may be deemed determined under Condition 34A (Deemed determination).

Condition 50 (Certifying payments)

Paragraph (3) has been amended to provide that, where Condition 48(2)(a) (Alternative B (Milestone

Payment Chart)) applies, certification in respect of the Design Fee and any other work Instructed under Condition 10B (Design process and Contract Sum) shall be monthly, under paragraph (2). During the initial period, while the Design Fee is being earned, there will normally be no achievement of Milestones.

Condition 62 (Subletting)

Reference to Condition 34A (Deemed determination) has been added to subparagraph (2)(a).

Condition 66 (Performance bond)

This Condition now provides for the performance bond to be delivered by the Contractor to the Employer within 21 Days of notice of possession being given under Condition 34 (Commencement and completion); or within 21 Days after the determination of the Contract Sum pursuant to Condition 10B (Design process and Contract Sum), whichever is later; rather than within 21 Days of acceptance of tender. This is because, while the Contractor's responsibility is limited to design (and perhaps some other preparatory work), it will hardly be appropriate to require a performance bond from him - it would be rather like requiring a performance bond from an architect or consulting engineer: and, until the Contract Sum has been determined, it will not be possible to calculate the amount of the performance bond. As pointed out in the commentary on Condition 34 (Commencement and completion), it is possible for the Employer to give notice of possession before the achievement of the Design Process Event (the definition of which includes determination of the Contract Sum).

ACKNOWLEDGEMENTS

The Government agency responsible for this document is -

Property Advisers to the Civil Estate (PACE)
Trevelyan House
Great Peter Street
London
SW1P 2BY

The development of this document was directed and greatly aided by a sub-group of the PACE Joint Users' Group ('JUG') of Government Departments, consisting of -

Roger Squires	PACE Central Advice Unit (Chairman)
Tony Amis	Highways Agency
Robert Boyd	Department of the Environment for Northern Ireland Construction Service
Charles Branch	Department of Trade and Industry
Mike Burt	HM Treasury (Procurement Practice & Development Group)
Ian Campbell	The Scottish Office
John Garnett	Consultant to the Ministry of Agriculture, Fisheries and Food
Sandra Hatfield	Defence Estates Organisation
Jeff Hogg	Lord Chancellor's Department
Lynne Jones	The Buying Agency
Gerry Marlow	Prison Service
Graham Mason	Prison Service
Robin McKelvey	Department of the Environment for Northern Ireland Construction Service
Deric McTier	Department of the Environment for Northern Ireland Construction Service
John Moore	Parliamentary Works Directorate
Robert Pilling	PACE Central Advice Unit
David Reid	Home Office
Richard Whittaker	English Heritage
Les Wilson	British Nuclear Fuels plc
Maureen Whyman	Inland Revenue

The substantive drafting and legal advice and assistance necessary for the preparation of this document was principally provided by Andrew Pike of -

Pinsent Curtis
Solicitors
3 Colmore Circus
Birmingham B4 6BH

Advice on Scots law was provided by David Henderson of -

MacRoberts
Solicitors
152 Bath Street
Glasgow G2 4TB

Advice on Northern Ireland law was provided by -

Departmental Solicitor's Office
Department of Finance & Personnel
Victoria Hall
12 May Street
Belfast BT1 4NL

Advice on risk and insurance matters was provided by Robin Keeling of -

Aon Group Limited
Construction & Engineering Division
East India House
109 - 117 Middlesex Street
London E1 7JF

Construction consultancy advice was provided by Brendan Murphy of -

Tarmac Services
37-39 Kew Foot Road
Richmond
Surrey TW9 2SS

Printed in the United Kingdom for The Stationery Office

J 85956 C20 7/99

ORDER FORM

Stationery Office publications are available from Stationery Office bookshops and agents and all other good booksellers, or direct:

By Mail order
Please complete this form and return it to The Stationery Office, The Publications Centre, PO Box 276, London, SW8 5DT.

By Telephone
Please ring 0870 600 5522 and quote your Mastercard/Visa/Amex/Diners/Connect card number or your Stationery Office account number.

By Fax
Please complete the credit card or the Stationery Office account section of the order form below and fax to 0870 600 5533

Online
Order direct from www.tso-online.co.uk

Please tick one of the following:

☐ Debit my Mastercard/Visa/Amex/Diners/Connect Card No.

Signature ______________________

Expiry date ______________________

☐ Charge to my Stationery Office Account No. ______________________

☐ I enclose a cheque for £ __________ made payable to 'The Stationery Office'

PLEASE COMPLETE IN BLOCK CAPITALS

Name ______________________

Address ______________________

____

____

____

Postcode __________ Tel/Fax ______________________

AEL

☐ Please tick this box if you do not consent to us using this information for the purpose of marketing products and services by *The Stationery Office* or other organisations.

Prices are correct at time of going to press but may be subject to change without notice.

Account holders should note that credit card transactions will not be shown on their statements.

The Stationery Office

RELATED PUBLICATIONS: THE GC/WORKS FAMILY OF CONTRACTS

Please send me the following publication(s):

Qty	Title	ISBN	Price
______	GC/Works/1 Contract for building and civil engineering major works (value over £150,000): GC/Works/1: Part 1: Contract for building and civil engineering major works, with quantities general conditions	0 11 702184 9	£20
______	GC/Works/1: Part 2: Contract for building and civil engineering major works, without quantities general conditions	0 11 702185 7	£20
______	GC/Works/1: Part 3: Contract for building and civil engineering major works, single stage design and build general conditions	0 11 702186 5	£20
______	GC/Works/1: Part 4: Contract for building and civil engineering major works, model forms and commentary	0 11 702187 3	£20
______	GC/Works/1: Complete set (Parts 1-4)	0 11 702188 1	£75
______	GC/Works/1: Part 5: Two stage design and build	0 11 702326 4	£20
______	GC/Works/1: Part 6: Construction Management	0 11 702327 2	£20
______	GC/Works/1: Part 7: Without quantities construction management trade contract	0 11 702401 5	£20
______	GC/Works/2: Contract for building and civil engineering minor works (value £25,000 - £200,000)	0 11 702152 0	£30
______	GC/Works/3: Contract for mechanical and electrical engineering works (any value)	0 11 702153 9	£30
______	GC/Works/4: Contract for building, civil engineering, mechanical and electrical small works (value up to £75,000).	0 11 702154 7	£30
______	GC/Works/5: Part 1: General conditions for the appointment of consultants	0 11 702310 8	£30
______	GC/Works/5: Part 2 (1999) : General conditions for the appointment of consultants – term contracts	0 11 702332 9	£30
______	GC/Works/6: Standard form of daywork contract *Available late 1999*	0 11 702328 0	£30
______	GC/Works/7: Standard form of measured term contract which is based on a schedule of rates *Available late 1999*	0 11 702329 9	£30
______	GC/Works/8: Specialist term contract for use where specified maintenance of equipment is required and can be costed per task *Available late 1999*	0 11 702330 2	£30
______	GC/Works/9: Lump sum maintenance contract (1-5 years) for operation, maintenance and repair of fixed mechanical and electrical plant, equipment and installations *Available late 1999*	0 11 702331 0	£30
	GC/Works Subcontracts – three new titles published by The Stationery Office in association with PACE and the Construction Confederation. Fully compatible with GC/Works and completely updated for the first time since 1985, this series is essential reading for subcontractors in construction and tendering for government work.		
______	Subcontract GW/S for use with GC/Works/1 Conditions *Available late 1999*	0 11 702340 X	£20
______	Articles of Agreement and Appendix *Available late 1999*	0 11 702341 8	£20
______	Subcontract GW/S for use with GC/Works/2 Conditions *Available late 1999*	0 11 702347 7	£20
______	Articles of Agreement and Appendix *Available late 1999*	0 11 702348 5	£20
______	Subcontract GW/S for use with GC/Works/3 Conditions *Available late 1999*	0 11 702349 3	£20
______	Articles of Agreement and Appendix *Available late 1999*	0 11 702350 7	£20
______	User's Guide to GW/S Subcontracts *Available late 1999*	0 11 702342 6	£20
______	PC/Works (1998) The conversion tool for the GC/Works range of contracts, as designed by the national law firm, Pinsent Curtis. A single user-friendly volume which allows you to translate the government forms of contract into a format which can be used for any construction project and for any employer.	0 11 702189 X	£50

Subtotal	**£**	
Handling	**£**	**2.94**
TOTAL	**£**	